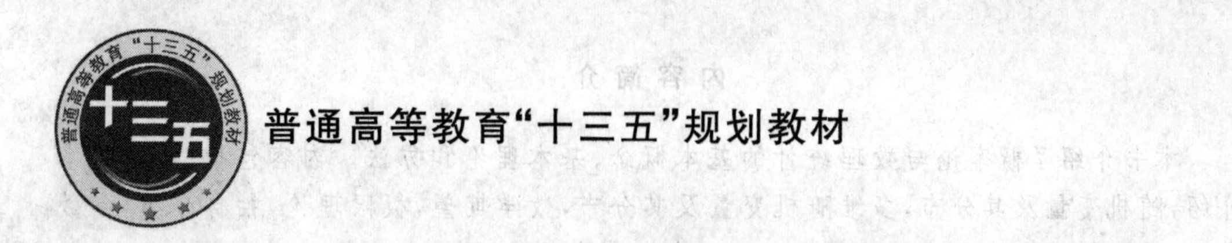

Probability Theory and Mathematical Statistics

概率论与数理统计

王学丽　杨建奎　郭永江　编

北京邮电大学出版社
www.buptpress.com

内容简介

本书介绍了概率论与数理统计的基本概念、基本理论和方法。内容包括：概率基本介绍，随机变量及其分布，多维随机变量及其分布，数学期望，极限理论，抽样分布，参数估计，假设检验和回归分析。每章课后配有丰富的习题，供学生练习之用。

本书是高校工科、理科（非数学专业）"概率论与数理统计"的双语教材，也可作为高等学校理工科各专业学生及教师的教材和参考书，还可供科技工作者阅读。

图书在版编目(CIP)数据

概率论与数理统计 = Probability Theory and Mathematical Statistics ：英文 / 王学丽，杨建奎，郭永江编. -- 北京 ：北京邮电大学出版社，2017.8

ISBN 978-7-5635-5169-9

Ⅰ. ①概… Ⅱ. ①王… ②杨… ③郭… Ⅲ. ①概率论—高等学校—教材—英文②数理统计—高等学校—教材—英文 Ⅳ. ①O21

中国版本图书馆 CIP 数据核字(2017)第 187356 号

书　　名：Probability Theory and Mathematical Statistics(概率论与数理统计)
著作责任者：王学丽　杨建奎　郭永江　编
责 任 编 辑：刘　颖
出 版 发 行：北京邮电大学出版社
社　　址：北京市海淀区西土城路 10 号(邮编：100876)
发 行 部：电话：010-62282185　传真：010-62283578
E-mail：publish@bupt. edu. cn
经　　销：各地新华书店
印　　刷：北京鑫丰华彩印有限公司
开　　本：787 mm×960 mm　1/16
印　　张：19.75
字　　数：320 千字
版　　次：2017 年 8 月第 1 版　2017 年 8 月第 1 次印刷

ISBN 978-7-5635-5169-9　　定　价：42.00 元

·如有印装质量问题请与北京邮电大学出版社发行部联系·

Preface

"Probability theory and statistics" has been an important mathematical course for years. It is a necessary project for the undergraduate students with a wide range of ability in science, engineering, management and business.

With the growing demands of studying abroad, bilingual education, as the preparation in advance, becomes increasingly urgent. This book is written for flattering this purpose. Of course, there have been already plenty of textbooks in English about this subject. However, used in a one-semester course, the existing textbooks with excessive contents obviously need to be adjusted. In this direction, we hope the publishing of this book could be a meaningful exploration.

Comparing with those traditional versions, we focus on the following aspects: (1) Under the framework of modern probability theory, axiomatic definitions of probability are systematically introduced. By sets we denote the random events, and further, by measurable functions (random variables) we simplify the form of describing the random phenomena into a mathematical way. Then the axioms get applications to measure. In this book, a lot of space is contributed to these important concepts of independence, dependent, conditional probability. (2) Following the probability measure setting, explicit expressions about the probability, such as distribution functions and probability density functions, are introduced with commonly used distributions as examples. (3) With the probability measures, we discover the characteristic features of random variables by expectation, variance, covariance and correlation. (4) For the trend of large amount of randomness, we introduce law of large number and central limit theorem as the end part of probability theory. (5) With the beginning of mathematical statistics, random samples and several important sampling distributions such as the chi-square (χ^2) distribution, the T distribution , the F distribution and the normal distribution et al, are given. (6) Learning the sampling distribution, we also represent the estimation of parameters, including point estimation and confidence interval estimation. (7)Furthermore, testing hypotheses as a part of statistical inference are illustrated by using some specific examples. (8) In the last portion, a simple linear statistical model called one linear regression model is demonstrated with several understandable examples. These Statistics parts contain a comprehensive but elementary survey of estimation, testing hypotheses, regression methods. The strengths and weaknesses of such basic concepts as maximum likelihood estimation, unbiased estimation, confidence intervals, levels of significance and regression idea are discussed from a contemporary viewpoint.

As a mathematical course, although everything turns to be theorems and formulas eventually, we hope the students can be well trained to build the intuition besides rigorous thinking. The students are also encouraged to modulate or verify the theoretical concepts through softwares, such as SAS, matlab and R et al, especially those in the statistics part. Unfortunately, we fail to involve the latter much in this book for space limit.

Chapter 1 to Chapter 2 are written by Yang Jiankui, Chapter 3 to Chapter 5, GuoYongjiang, and Chapter 6 to Chapter 9, Wang Xueli. The authors also want to represent the sincere thanks for the guidance of the colleagues and the financial support from Beijing University of Posts and Telecommunications.

Limited to the level of authors, deficiencies and errors of this book are inevitable. We hope the criticism and corrections, truly.

目　录

Chapter 1 Introduction to Probability

1.1 Introduction

Uncertainty was born with the universe. As R. Deep said, "there is even evidence from quantum physics that suggests that God does indeed play dice with the universe, at least at the micro-infrastructure." See [1].

For human beings, a random phenomenon is a situation in which we know what outcomes could happen, but we do not know which particular outcome did or will happen. We deal with many random phenomena every day. For example, you can not predict which point appears before rolling a die, although you know all possible results. Coins flipped into the air will show the heads or the tails you can never know unless it is steady. Is it a sunny day or cloudy one when we get up every day morning? Maybe we can predict if it is raining or not tomorrow correctly for most of time. However, to predict the probability of raining is more responsible. Does the stock price will go upwards or downwards? What is the variability, such as how high or how low does it go? We need to manage these things carefully to get paid well.

As we know, some of the random phenomena play important roles in our lives really. However, it seems not elegant that chance phenomena got people's notice. In ancient Egypt tombs, dice (called astragal) were found. Coincidentally, they also appeared in Mesopotamian and Indian civilization more than 5 000 years ago. People managed the chances to win the interest in gambling. In fact, what promoted the growth of the real probability theory is the developing needs of the economy in the earliest date back to the fifteenth century, when is the beginning of the development of international trade. In the long voyage oversea, some valuable cargo insurance was needed. So when the occurrence of loss of goods was specified in the insurance contract, the owner can be compensated. Then risk assessment and calculations of premiums become formal.

It is a long time before the concept of probability was defined under the framework of the axiomatic system. Most recent theoretical discussion was started by the sixteenth century Italian mathematician Cardano, and later by the French mathematicians Pascal and Fermat through letters in 1654. The communications on this issue stimulated many European mathematicians to explore similar problems. Few years later, the Dutch physicist Huygens (1629-1695) published his formal writings and firstly defined the concept of expectation. Then after Jakob Bernoulli (1654-1705) and other mathematicians (Abraham de Moivre (1667-1754), Pierre Reymond Montmort (1678-1719), Thomas Bayes (? -1761), George Louis Buffon (1707-1788), Daniel Bernoulli (1700-1782), Joseph Louis Lagrange (1736-1813)) completed in the eighteenth century about classical probability theory.

Then probability had been widely applied in other disciplines, which promoted the probability theory itself. For example, the least square method (LSM) was independently given by Adrien-Marie Legendre (1752-1833) and Karl Friedrich Gauss (1777-1855) in 1806 and 1809 because of the need in astronomy and physics measurements.

Pierre-Simon Laplace (1749-1827) published his greatest achievement in terms of probability theory, namely "The Central Limit Theorem" in 1810, which is an important theoretical basis of modern probability theory and statistics. His book " Thorie Analytique des Probabilits" explicitly proposes the classical definition of probability.

Due to the research and development of "measure theory", probability theory axiomatic system was established. Some elementary probability theories and definitions unable to be explained before can be interpreted by the axiomatic language. This culminated in modern probability theory. The foundation work mainly was established by Andrey Nikolaevich Kolmogorov (1903-1987) in 1933. Then the probability theory was modernized and turned to be a strict and independent science discipline.

The seventeenth century is also when the sister discipline of probability, statistics began to sprout. To conscription or taxation, governments in Europe began collecting such as birth, death, marriage and other demographic information. There are applications of statistics in the commercial insurance and actuarial industry. We do not dig deeply the statistics history for now but postpone it to the statistics part in this book.

1.2 Interpretations of Probability

There are usually two viewpoints of probability: subjective and objective. For those subjectivists, they determine the probability of some event individually through their own knowledge and evaluation. For example, on the occasion of the flip of a coin to guess the probability of the head, the subjectivists will make their decision according to the evaluation whether the coin is fair. Then they will have a little trouble with randomness because of the lack of a pattern which concludes the probability. For those who favor the objective, they would like to make a series of trials of the coin flip experiment. They toss the coin 20 times in the similar circumstances, more or less, and then calculate the frequency of the head. The frequency is claimed to be the probability of the head. Of course, both subjectivists and objectivists may have the same conclusion. However, the objectivists see a pattern and take the advantage obviously.

Some will doubt if the frequency is indeed the probability. Then we arrive at the measure problem. In the history, there are different viewpoints of the measure: classical priori (equally likely outcomes) and empirical or relative frequency (posteriori). Neither of the viewpoints is rigorous. The equally likely outcomes do not fit all circumstances. Furthermore, the paradox will arise even it seems reasonable to assume this priori in some cases, such as the famous Betrand's Paradox in Example 1. 6. 3 (Interested readers refer to [3]). As to the latter, the frequency is dependent of the time of trials, experiment conditions, the external environment and so on. Different frequencies will probably be achieved in different experiments. Fortunately, in the latter part of the book, we can see that the frequency will approximately converge to the probability with the increasing of the trial times.

The following table contains some data of the coin flip experiment by several scientists since the 18th century. The frequency of the head is close to 0. 5 with large amount of trial times.

Table 1. 2. 1

Experimenter	Trial times	Times of the head	Frequency
Comte de Buffon	4 040	2 048	0. 506 9
De Morgan	4 092	2 048	0. 500 5
William Feller	10 000	4 979	0. 497 9
Karl Pearson	24 000	12 012	0. 500 5
Romanovsky	80 640	39 699	0. 492 3

Here we introduce some technical words to complete this section. An **experiment** is a repeatable and stochastic process with a sequence of repeatable and stochastic trials. The result of each experiment is called an **outcome**, which is sometimes called **sample** (**point**). The experiment usually results in many outcomes. We already know all possible outcomes without performing the experiment, but do not know which one will be obtained in some experiment. Any combination of outcomes of specific interest is an **event.** Specially, an event is called a **simple** event if it consists of only one outcome, a multiple event otherwise. The collection (set) of all outcomes constitutes a **sample space.**

Let's see an example for these words.

Example 1.2.1 (Experiments)

There are 5 experiments:

(1) Toss a coin 2 times to observe the head or the tail;

(2) Observe sunny, windy, cloudy, rainy days in a month;

(3) Call times for a call center;

(4) Choose one from 10 000 inspected transistors to check it is of high or low quality;

(5) Record the air temperature at a certain location every day at noon for 90 successive days.

From (1) to (5), it is easy to see that the trials in these processes are all repeatable and stochastic. So they are all experiments.

In (1), we know all possible results of every 2 tosses are {H,H},{H,T},{T,T} and {T,H}, where H represents head and T tail.

The four results are all outcomes of the experiment and constitute the sample space. The event we have at least one head is a combination of outcomes {H,H},{H,T} and {T,H}. The event we have no tails is just a simple event consisting of only one outcome {H,H}.

The rest parts (2)-(5) can be analyzed similarly and are left to the students.

Through the example, it is easy to see that each outcome can be regarded as a point, or an element, in the sample space. The event is just a subset of the sample space. Because of this interpretation, the language and concepts of set theory become natural. It is convenient to denote experiments, events and sample space by sets. Hence, before we measure the probability of a random event, we involve some set expressions to describe the events and reveal the relationship between them.

1.3 Set Algebra

The basic ideas and notations of set theory now are reviewed.

We usually denote by S the sample space of an experiment. Capital letters designate events, such as A, B, $\cdots$. These events are subsets of sample space. Small letters or numbers usually designate the element of sample space, i. e. , the possible sample, such as $a,b,\cdots$.

We say that some event A has occurred in the experiment, when the outcome is some element a belonging to A, i. e. , the outcome $a\in A$.

Example 1.3.1 Roll the Six-sided Die

All possible outcomes are 1,2,3,4,5 and 6. We can simply use $S=\{1,2,3,4,5,6\}$ to denote the sample space. $1,2,3,4,5,6\in S$. When we say the event A that we have an even number occurred, it means the outcome of this experiment is one of 2,4 and 6. So $A=\{2,4,6\}$. Similarly, that event $B=\{1,3,5\}$ occured means we have an odd number. If event $C=\{1,2,3\}$ occurred, then the number we got is no greater than 3. The event that we have a number no greater than 6 is just the sample space S. S is deterministic to occur in every trial because it contains all outcomes. Such an event S is also called a **certain event.** On the contrary, empty set $\phi=S^c$ should be called an **impossible event** because it contains no samples, where S^c is the complement of S. For example, event D means we have a number greater than 6. Then $D=\phi$. It is definitely not going to occur. Obviously $A,B,C,D\subset S$.

Then we can give new meanings to these set relations.

Complement: $A^c=\{s\in S:s\notin A\}$.

Subset: $A\subset B$ if and only if $\forall s\in A$, then $s\in B$.

Union: $A\cup B=\{s\in S:s\in A \text{ or } s\in B\}$.

Intersection: $A \cap B = \{s \in S : s \in A \text{ and } s \in B\}$.

Difference: $A - B = \{s \in S : s \in A \text{ and } s \notin B\}$

In an experiment, if event A occurs then A^c can not occur certainly. We call A^c the complement event of A. If $A \subset B$, then that event A occurs implies event B occurs. That the union event $A \cup B$ occurs implies at least one of A and B occurs. That the intersection event $A \cap B$ happens means both A and B happen. Sometimes $A \cap B$ can be written as AB for simplicity. Finally, that the difference event $A - B$ occurs means B does not occur but A does.

Example 1. 3. 2

With the same setting in Example 1. 3. 1,

(1) represent the event we have an odd number without using B.

(2) find the event that we have a number less than 3 if $E = \{3\}$.

(3) find the event that we have an odd number except 5.

Solution.

(1) We have an odd number, which is equivalent that we can not have an even number. The answer is A^c; (2) C occurs and E does not, that is $C - E$; (3) Both B and C occur, that is, $B \cap C$.

Example 1. 3. 3

There is a bulb in the circuit as shown in Figure 1. 3. 1.

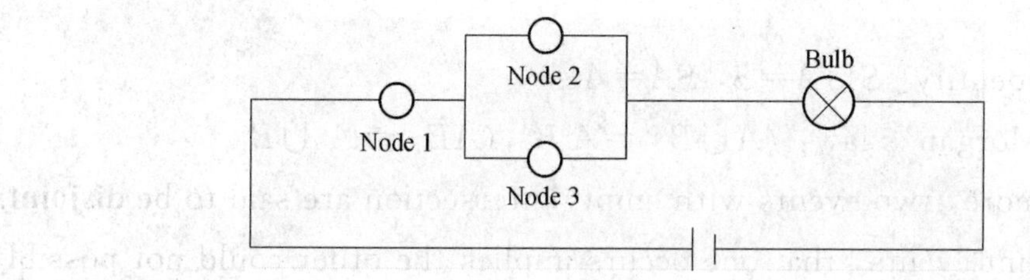

Figure 1. 3. 1

There are also three nodes 1, 2 and 3. We use a 3-dimension vector to denote the status from Node 1 to Node 3, with element 1 representing that a node is connected and 0 otherwise. Obviously, all possible cases are contained in

$S=\{(1,0,0),(1,1,0),(1,0,1),(1,1,1),(0,0,0),(0,1,0),(0,0,1),(0,1,1)\}$.

We use $a_1,\cdots,a_8$ to denote the 8 elements in S respectively. Find

(1) event A: Node 1 is connected.

(2) event B: Node 2 is connected.

(3) event C: Node 3 is connected.

(4) event D: the bulb is light.

Solution.

(1) $A=\{a_1,a_2,a_3,a_4\}$.

(2) $B=\{a_2,a_4,a_6,a_8\}$.

(3) $C=\{a_3,a_4,a_7,a_8\}$.

(4) To keep the bulb lighting, we need Node 1 is connected and at least one of Node 2 and Node 3 is connected. That is, $D=A\cap(B\cup C)$.

In the rest part of the book, we denote by S the sample space without special explanation. When we use the set algebra to describe the events, the following laws still hold for the events S,A,B,C.

(1) Commutative law: $A\cup B=B\cup A$, $A\cap B=B\cap A$.

(2) Distributive law: $A\cup(BC)=(A\cup B)(A\cup C)$, $A(B\cup C)=(AB)\cup(AC)$.

(3) Associative law: $A\cup(B\cup C)=(A\cup B)\cup C=A\cup B\cup C$, $A(BC)=(AB)C$.

(4) Set identity: $S\cup A=S$, $SA=A$.

(5) De Morgan's law: $(A\cup B)^c=A^cB^c$, $(AB)^c=A^c\cup B^c$.

Furthermore, two events with empty intersection are said to be **disjoint.** For two disjoint events, that one occurs implies the other could not possibly have occurred. A sequence of events, among which any two events have an empty intersection, are called **mutually exclusive**, meaning that when one event occurs, the rest events could not possibly have occurred.

1.4 The Definition of Probability

Now we start to define the probability of the events. The infrastructure is based on three probability axioms. Then everything follows. Let $P(A)$ indicate the probability that event A will occur with $A \subset S$. Recall the three axioms:

Axiom 1. For every event A, $P(A) \geqslant 0$.

Axiom 2. $P(S)=1$.

Axiom 3. For every infinite sequence of mutually exclusive events A_1, $A_2, A_3, \cdots$, $P(\bigcup_{n=1}^{\infty} A_n) = \sum_{n=1}^{\infty} P(A_n)$.

The 3rd one is also called **additive** property.

We can easily have some corollaries from the axioms.

Theorem 1.4.1 $P(\phi)=0$.

Intuitively, the chance of the impossible event occurring should be 0. Now we prove it with acknowledging the axioms.

Proof.

From Axiom 1, $P(\phi) \geqslant 0$. There is a sequence of empty sets $\phi_n = \phi, n = 1, 2, \cdots$, which are mutually exclusive. It is easy to see that $\bigcup_{n=1}^{\infty} \phi_n = \phi$. From Axiom 3, we have

$$P(\phi) = P(\bigcup_{n=1}^{\infty} \phi_n) = \sum_{n=1}^{\infty} P(\phi_n) = \sum_{n=1}^{\infty} P(\phi),$$

which implies $P(\phi)=0$.

Theorem 1.4.2

For every finite sequence of n mutually exclusive events $A_1, A_2, A_3, \cdots$, A_n, $P(\bigcup_{i=1}^{n} A_i) = \sum_{i=1}^{n} P(A_i)$.

Proof. Define $\phi = A_{n+1} = A_{n+2} = \cdots$. The theorem is immediate from

Axiom 3 and Theorem 1.4.1.

Theorem 1.4.3

Assume two events $A, B \subset S$.

(1) $P(A)=1-P(A^c)$.

(2) $P(A)=P(AB)+P(AB^c), P(B)=P(AB)+P(A^cB)$.

(3) If $A \subset B$, then $P(A) \leqslant P(B)$ and $P(B-A)=P(B)-P(A)$.

Proof.

$A \cup A^c = S, AA^c = \phi$. From Axiom 2 and Theorem 1.4.2, we have

$$1=P(A \cup A^c)=P(A)+P(A^c).$$

Then (1) is immediate.

For (2), $A=AB \cup AB^c$ and $AB \cap AB^c=\phi$. From Theorem 1.4.2, we have $P(A)=P(AB \cup AB^c)$. The other one is immediate from the symmetry.

For (3), if $A \subset B$, then $B=(B-A) \cup A$.

Further,

$$P(B)=P(B-A)+P(A),$$

which is equivalent to

$$P(B-A)=P(B)-P(A).$$

From Axiom 1, $P(B-A) \geqslant 0$. Then $P(A) \leqslant P(B)$.

Theorem 1.4.4 Sum Formula

(1) $P(A \cup B)=P(A)+P(B)-P(AB)$.

(2)
$$P(A \cup B \cup C)=P(A)+P(B)+P(C) -P(AB)-P(BC)-P(AC)+P(ABC)$$

(3) Generally, for a sequence of events $A_1, A_2, A_3, \cdots, A_n$,

$$P(\bigcup_{i=1}^{n} A_i)=\sum_{i=1}^{n} P(A_i)-\sum_{1 \leqslant i<j \leqslant n} P(A_iA_j) + \sum_{1 \leqslant i<j<k \leqslant n} P(A_iA_jA_k)-\cdots + (-1)^{n-1} P(A_1A_2\cdots A_n)$$

Proof.

$A \cup B=A \cup (B-AB)$, $A(B-AB)=\phi$.

Then $P(A \cup B)=P(A \cup (B-AB))=P(A)+P(B-AB)$. (1.4.1)

Since $AB \subset B$, from (3) in Theorem 1.4.3,

$$P(B-AB)=P(B)-P(AB). \quad (1.4.2)$$

Substituting (1.4.2) into (1.4.1), we have (1).

There is a Venn diagram in the following remark for some intuitive senses.

For (2) and (3), we do not show the proof here for the complexity. Readers interested can try it by the induction.

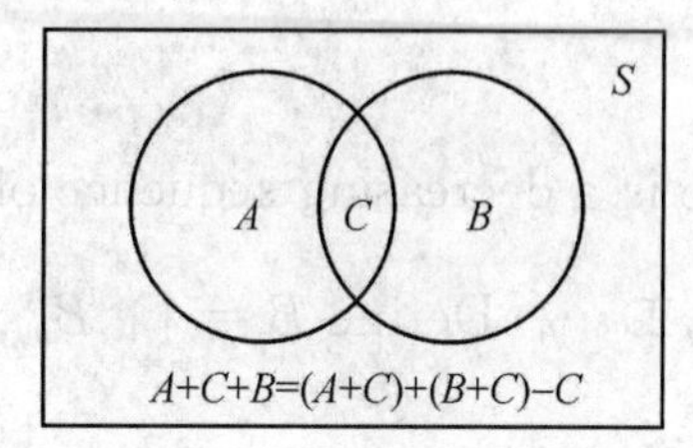

Figure 1.4.1

Remark: Venn Diagram Illustrating

Example 1.4.1 Diagnosing Diseases

A patient arrives at a doctor's office with a sore throat and low-grade fever. After an exam, the doctor concludes that the patient has either a bacterial infection or a viral infection or both. She decides that there is a probability of 0.7 that the patient has a bacterial infection and there is a probability of 0.5 that the person has viral infection. What is the probability that the patient has both infections?

Solution.

B: bacterial infection. V: viral infection. Since the patient has either a bacterial infection or a viral infection or both for sure. $S=B \cup V$. We have already know that

$$P(B)=0.7, \quad P(V)=0.5.$$

What we want to know is $P(BV)$.

$$P(BV)=P(B)+P(V)-P(B \cup V)=0.7+0.5-1=0.2.$$

Example 1.4.2

Two events $A, B \subset S$ with probability $P(A)=0.8$ and $P(A-B)=0.4$.

Find $P((AB)^c)$.

Solution.

$$P(AB)=P(A-(A-B))=P(A)-P(A-B)=0.4.$$

Then $P((AB)^c)=1-0.4=0.6$.

Theorem 1.4.5 The Continuity of Probability

(1) Suppose that there is an increasing sequence of events $A_1, A_2, \cdots, A_n, \cdots$. That is, $A_1 \subset A_2 \subset \cdots \subset A_n \subset \cdots$. Define $A = \bigcup_{n=1}^{\infty} A_n$. Then we have $P(A)=\lim_{n\to\infty} P(A_n)$.

(2) Suppose that there is a decreasing sequence of events $B_1, B_2, \cdots, B_n, \cdots$. That is, $B_1 \supset B_2 \supset \cdots \supset B_n \supset \cdots$. Define $B = \bigcap_{n=1}^{\infty} B_n$. Then we have $P(B)=\lim_{n\to\infty} P(B_n)$.

Proof.

Here we only prove (1). The proof of (2) can also be proved by following similar steps.

Define a sequence of events C_n as follows:

$$C_1=A_1, C_2=A_2-A_1, \cdots, C_n=A_n-A_{n-1}, \cdots$$

Then $C_n, n=1,2,\cdots$ are mutually exclusive and $A=\bigcup_{n=1}^{\infty} A_n = \bigcup_{n=1}^{\infty} C_n$. From Axiom 3,

$$P(A)=P(\bigcup_{n=1}^{\infty} C_n)=\sum_{n=1}^{\infty} P(C_n)=\sum_{k=1}^{n} P(C_k)$$

$$=\lim_{n\to\infty}\sum_{k=2}^{n}[P(A_k)-P(A_{k-1})]+P(A_1)=\lim_{n\to\infty} P(A_n).$$

1.5 Finite Sample Spaces

This section concerns with a kind of probability model, which involves finite samples. As the name suggests, the characteristic of this model is that the sample space consists of finite samples.

Assume that $S=\{s_1, s_2, \cdots, s_n\}$ with probability equipment $p_1, p_2, \cdots, p_n$ satisfies the following three conditions:

(1) $p_i>0, i=1,2,\cdots,n.$

(2) $\sum_{i=1}^{n} p_i = 1.$

(3) $P(\{s_i\})=p_i.$

Then it is easy to see that the probability of each event A can then be found by adding the probabilities p_i of all outcomes s_i that belong to A.

There is a special case of the above model, which is called a **simple sample space.**

Definition 1.5.1

A sample space $S=\{s_1, s_2, \cdots, s_n\}$ is called a simple sample space if the probability assigned to each of the outcomes $s_1, s_2, \cdots, s_n$ is $\frac{1}{n}$.

We can see that this simple sample space consists of equally likely outcomes. If an event A in this simple sample space contains exactly m outcomes, then

$$P(A)=\frac{m}{n}=\frac{\# A}{\# S}.$$

Here, the symbol "$\# A$" stands for the number of elements in A.

Example 1.5.1 Rolling Two Dice

Consider an experiment in which two balanced dice are rolled, and calculate the probabilities of the possible values of the two numbers that may appear.

For simplicity we assume that the two dice are distinguishable. The fact that two dice are balanced implies that each pair of numbers is equally likely. Of course, there are only finite outcomes:

(1, 1) (1, 2) (1, 3) (1, 4) (1, 5) (1, 6)

(2, 1) (2, 2) (2, 3) (2, 4) (2, 5) (2, 6)

(3, 1) (3, 2) (3, 3) (3, 4) (3, 5) (3, 6)

(4, 1) (4, 2) (4, 3) (4, 4) (4, 5) (4, 6)

(5, 1) (5, 2) (5, 3) (5, 4) (5, 5) (5, 6)

(6, 1) (6, 2) (6, 3) (6, 4) (6, 5) (6, 6)

Here, numbers (x, y) represents the pair of numbers, and x is the number that appears on the first die and y is the number that appears on the second one. Therefore, S comprises 36 outcomes and is a simple sample space. The probability of each simple event is $\frac{1}{36}$. Readers interested may consider the scenario that we do not distinguish the two dice.

We define that p_i is the probability that the sum of the pair of numbers is i, i. e. , $p_i \doteq P(\{(x,y) \in S : x+y=i\})$. All possible values of i are 2, 3, ⋯, 12. Find p_2, p_7 and p_6.

Solution.

$$p_2 = P(\{(1,1)\}) = \frac{1}{36}.$$

$$p_7 = P(\{(1,6),(6,1),(2,5),(5,2),(3,4),(4,3)\}) = \frac{6}{36} = \frac{1}{6}.$$

$$p_6 = P(\{(1,5),(5,1),(2,4),(4,2),(3,3)\}) = \frac{5}{36}.$$

From the above discussion we conclude that the key to calculate the probability in the simple space is to count the number of outcomes of some event. The combinatorial method is a powerful tool to count numbers.

Here we slightly recall combinations. In this book, we use C_n^k to define the number of combinations of n elements taken k at a time, P_n^k the number of permutations of n elements taken k at a time.

There are some simple formulae:

(1) $C_n^k = \frac{n!}{k!\,(n-k)!}$, $P_n^k = \frac{n!}{(n-k)!}$, $P_n^k = C_n^k \cdot k!$.

(2) (Binomial theorem) For all numbers x and y and each positive integer n, $(x+y)^n = \sum_{k=0}^{n} C_n^k x^k y^{n-k}$.

(3) $\sum_{k=0}^{n} C_n^k = 2^n$, $\sum_{k=0}^{n} (-1)^k C_n^k = 0$.

Example 1.5.2 Selecting a Committee

Suppose that a committee composed of 3 students is to be selected randomly from a class of 20 students. Find the probability that Luren Jia (one student in this class) is selected.

Solution.

3 students are randomly selected, which implies that all combinations of 3 students are equally likely. The problem turns to be a simple sample space model.

Notice that the total number of different groups of students that might be on the committee is

$$C_{20}^{3}=\frac{20!}{3!17!}.$$

The total number of committees with Luren Jia is

$$C_{19}^{2}=\frac{19!}{2!17!}.$$

Then the probability that Luren Jia is selected is

$$p=\frac{C_{19}^{2}}{C_{20}^{3}}=\frac{3}{20}.$$

Example 1.5.3

There are five cards with numbers 1, 2, 3, 4, 5 respectively. Select randomly 3 cards one by one and place them from the left side to the right side. Then we have a new three-digit number. Find the probability that the new number is an even one.

Solution.

There are P_5^3 numbers. The new number is even, which means the last digit should be 2 or 4. Therefore, there are $C_2^1P_4^2$ even numbers. Then the probability we need is

$$p=\frac{C_2^1P_4^2}{P_5^3}=\frac{2}{5}.$$

Now let's see another upgraded example.

Example 1.5.4

Keep the settings in Example 1.5.3 unchanged except that the numbers on the five cards are 1, 2, 3, 4, 4. We still want to know the probability that the new three-digit number is an even one.

Solution.

In fact, there are still P_5^3 combinations of 3 cards although there are numbers repeated. The new number is even, which means the last digit should be 2 or 4. Again, there are $C_3^1 P_4^2$ combinations with the new number being even numbers. Then the probability we need is

$$p=\frac{C_3^1 P_4^2}{P_5^3}=\frac{3}{5}.$$

Example 1.5.5 Sampling without Replacement

Suppose that there are 15 new and 30 used ping-pang balls, and that 10 balls are taken out for the match. We shall determine the probability p that exactly 3 new balls will be selected.

Solution.

It is easy to see that

$$p=\frac{C_{15}^3 C_{30}^7}{C_{45}^{10}}.$$

Continue this example. We put back the balls when the match is over. If we choose 10 balls again for another match, what is the probability p that exactly 3 new balls will be selected. The new combination seems to be depending on one in the former match. For this, we postpone to Example 1.7.5.

Example 1.5.6 Matching Problem

There are 3 tables labeled by numbers 1, 2, 3. Three cards marked with 1, 2, 3 are put onto the tables, one card on one table. If the card number is the same as the table number, we call it a match. We shall determine the probability that we have at least one match. Furthermore, if we have 100 tables and 100 cards with numbers 1, 2, …, 100, what is the probability that we have at least one match.

Solution.

For the former case, there are 3! permutations for the three cards. The total number of only one match is 3. The total number of only two matches (three matches at the same time) is 1. Then the probability that we have at least one match is $\frac{2}{3}$.

For the latter case, it seems tough. The sum formula works here. Define that A_i is the event that the card with number i matches the ith table. Then what we need is $P(\bigcup_{i=1}^{100} A_i)$, to find which, we make the following preparation.

$$P(A_1)=\frac{99!}{100!}, P(A_2)=\frac{99!}{100!}, \cdots, P(A_{100})=\frac{99!}{100!},$$

$$P(A_1A_2)=\frac{98!}{100!}, P(A_2A_3)=\frac{98!}{100!}, \cdots, P(A_iA_j)=\frac{98!}{100!}, \cdots,$$

$$P(A_{99}A_{100})=\frac{98!}{100!},$$

$$P(A_iA_jA_k)=\frac{97!}{100!}, \cdots, P(A_1A_2\cdots A_{100})=\frac{1}{100!}.$$

Then

$$\begin{aligned}P(\bigcup_{i=1}^{100} A_i) &= \sum_{i=1}^{100} P(A_i) - \sum_{1\leqslant i<j\leqslant 100} P(A_iA_j) + \sum_{1\leqslant i<j<k\leqslant 100} P(A_iA_jA_k) + \cdots + (-1)^{99} P(A_1A_2\cdots A_{100}) \\ &= 100\times\frac{1}{100} - C_{100}^2\times\frac{1}{100\times 99} + C_{100}^3\times\frac{1}{100\times 99\times 98} + \cdots + (-1)^{99}\frac{1}{100!} \\ &= 1 - \frac{1}{2!} + \frac{1}{3!} + \cdots + (-1)^{99}\frac{1}{100!} = \sum_{i=1}^{100}(-1)^{i-1}\frac{1}{i!}.\end{aligned}$$

1.6 Geometry Probability Setting

As we know that the main characteristic of simple sample space contains two parts: finite samples and equally likely outcomes. The probability model in this section inherits the latter part and expands the space of finite samples into one of infinite samples.

Since the events usually contain infinite samples, we cannot use the ratio of the sample numbers in events to ones in the sample space. Then we use the geometrical measure of the event instead of the sample number.

Example 1.6.1 Random Number

We select a real number from the interval $[0,1]$ at random. Find (1) the probability that the number is 0.5, and (2) the probability that the number is less than 0.5.

Solution.

The sample space is $S=\{x\in[0,1]\}$. There are infinite samples and all samples are equally likely chosen.

(1) $A=\{0.5\}$, $P(A)=\dfrac{\text{measure of } A}{\text{measure of } S}=\dfrac{0}{1}=0$,

(2) $B=\{x\in[0,0.5)\}$, $P(B)=\dfrac{\text{measure of } B}{\text{measure of } S}=\dfrac{0.5}{1}=0.5$,

where we mean the length by "measure".

Remark 1.6.1

We call ϕ impossible event and $P(\phi)=0$. Does any event with zero probability have to be an impossible event? The answer is definitely NO from above example.

Example 1.6.2 Meeting Problem

Luren Jia and Luren Yi will go to some cafe to have a date. Luren Jia will arrive at a random moment from 3 p.m. to 5 p.m., Luren Yi from 4 p.m. to 6 p.m. They shall arrive independently and the one arriving earlier will leave after 10 minutes if the other one fails to make it. Determine the probability that they really have a date.

Solution.

The sample space is $S=\{(x,y)\in[3,5]\times[4,6]\}$, where x is the arriving time for Luren Jia and y for Luren Yi.

The event they really have a date is

$$A=\left\{(x,y)\in S: |x-y|\leqslant\frac{1}{6}\right\}.$$

Then

$$P(A)=\frac{\text{aera of } A}{\text{aera of } S}=\frac{1/3}{4}=\frac{1}{12}.$$

This method seems powerful here. However, it also brings us confusion. That is, it is difficult to figure out the equally likely outcomes.

Example 1.6.3 Bertrand's Paradox

Consider an equilateral triangle inscribed in a unit circle. Suppose that a chord of the circle is chosen at random. What is the probability that the chord is longer than a side of the triangle?

Bertrand gave three different arguments. All of them seem valid, yet yielding different results.

(1) The "random endpoints" method: Start to draw a chord from one vertex of the triangle with the ending point of the chord randomly chosen on the circumference of the circle. Observe Figure 1.6.1 that if the other chord endpoint lies on the arc between A and B, the chord is longer than a side of the triangle. The length of the arc is one third of the circumference of the circle, therefore the probability that a random chord is longer than a side of the inscribed triangle is 1/3.

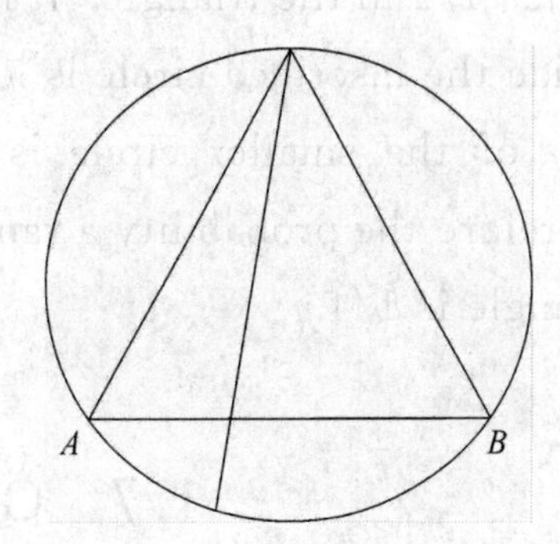

Figure 1.6.1

(2) The "random radius" method: Observe Figure 1.6.2. Choose a radius of the circle perpendicular to the two chords AB and CD with two intersection points E and F, and construct the chords perpendicular to the radius and between AB and CD, which are all longer than a side of the triangle. That is, the intersection points of those chords and the radius are between E and F. The length of the line EF is 1. Then the probability a random chord is longer than a side of the inscribed triangle is 1/2.

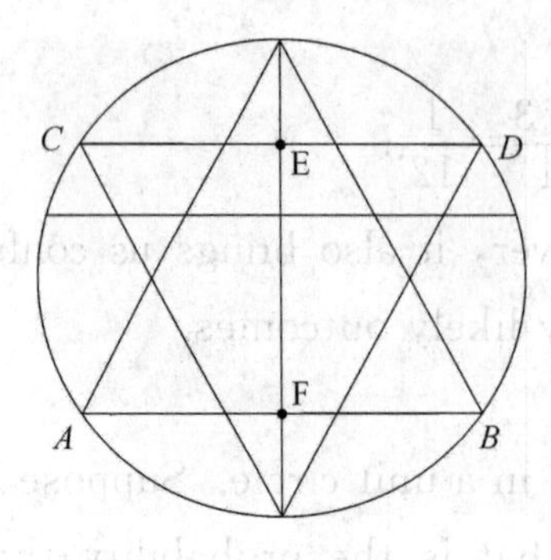

Figure 1.6.2

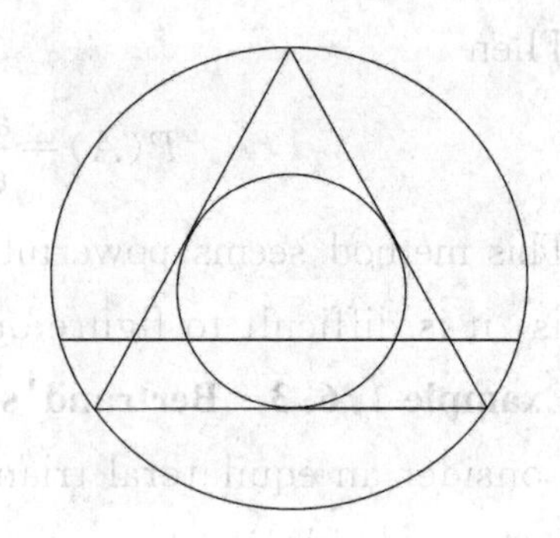
Figure 1.6.3

(3) The "random midpoint" method: Construct an inscribed circle of radius 1/2 in the triangle. It is easy to prove that any chord with its midpoint inside the inscribed circle is longer than a side of the inscribed triangle. The area of the smaller circle is one fourth of the area of the larger circle, therefore the probability a random chord is longer than a side of the inscribed triangle is 1/4.

1.7 Conditional Probability

Sometimes the possibility that an event occurs will change with updating information or other related events occurring. In this section, we shall now study the way in which the probability of an event A changes if it has already been learned that some other event B has occurred. This new probability of A is called the conditional probability of the event A given that the event B has occurred, which is denoted by $P(A \mid B)$. For convenience, we usually read $P(A|B)$ as the conditional probability of A given B.

How do we measure this new probability? Let's see an example to get some intuitions to the detailed definition.

Example 1.7.1 Rolling a Die

Suppose that one balanced die was rolled and we observed the number was less than 5. This event is denoted by A. Then

$$P(A)=\frac{\# A}{\# S}=\frac{4}{6}=\frac{2}{3}.$$

Furthermore, we denote by B the event that the number was odd. Then

$$P(B)=\frac{\# B}{\# S}=\frac{3}{6}=\frac{1}{2},$$

$$P(AB)=\frac{\# AB}{\# S}=\frac{2}{6}=\frac{1}{3}.$$

Now given B, the probability of A intuitively is

$$P(A|B)=\frac{\#\{1,3\}}{\#\{1,3,5\}}=\frac{\# AB}{\# B}=\frac{\# AB/\# S}{\# B/\# S}=\frac{P(AB)}{P(B)}=\frac{2}{3}.$$

This example shows us clues how we define a conditional probability. For concreteness, the denominator $P(B)$ should be positive.

Definition 1.7.1

There are two events A and B with $P(B)>0$. The probability of A given B is $P(A|B)=\dfrac{P(AB)}{P(B)}$.

Example 1.7.2

There are three events A_1, A_2 and B with $A_1A_2=\varnothing$, $P(B)>0$. Prove that $P(A_1\cup A_2|B)=P(A_1|B)+P(A_2|B)$.

Proof.

$$P(A_1\cup A_2|B)=\frac{P((A_1\cup A_2)B)}{P(B)}=\frac{P(A_1B\cup A_2B)}{P(B)}$$

$$=\frac{P(A_1B)}{P(B)}+\frac{P(A_2B)}{P(B)}=P(A_1|B)+P(A_2|B).$$

Example 1.7.3

Suppose that the probability that some kind of animals can live for more than 20 years is 0.8, 30 years 0.3. There is one of this kind of animals that has been living for more than 20 years. Find the probability that it can live for more than 30 years.

Solution.

A: It can live for more than 30 years.

B: It has been living for more than 20 years.

Notice that $A \subset B$ since that it can live for more than 30 years implies it has been living for more than 20 years.

The probability we want is $P(A|B)=\dfrac{P(AB)}{P(B)}=\dfrac{P(A)}{P(B)}=\dfrac{3}{8}$.

Theorem 1.7.1 The Multiplication Rule for Conditional Probabilities

There are two events A and B with $P(B)>0$. Then we have

$$P(AB)=P(B)P(A|B). \tag{1.7.1}$$

Further, assume that there is a sequence of events $A_1, A_2, \cdots, A_n$ which satisfy $P(A_1)>0, P(A_1A_2)>0, \cdots, P(A_1A_2\cdots A_{n-1})>0$.

Then

$$P(A_1A_2\cdots A_n)=P(A_1)P(A_2|A_1)P(A_3|A_1A_2)\cdots P(A_n|A_1A_2\cdots A_{n-1}). \tag{1.7.2}$$

Proof.

Formula (1.7.1) is direct from the definition of conditional probability. Formula (1.7.2) is simple by involving induction.

This theorem is very helpful for the probability of the intersection of a sequence of events.

Example 1.7.4 Principle of Drawing Lots

There is one movie ticket for 5 persons. They determine who will win the ticket through drawing 5 lots one by one. Is the probability that the person who draws the first will get the ticket larger than the probability that the other persons will get the ticket?

Solution.

A_i: The person who draws the ith lot gets the ticket.

$$P(A_1)=\frac{1}{5}.$$

Notice that $A_2 \subset A_1^c$. Then

$$P(A_2)=P(A_2A_1^c)=P(A_1^c)P(A_2|A_1^c)=\left(1-\frac{1}{5}\right)\times\frac{1}{4}=\frac{1}{5}.$$

Similarly, $A_3 \subset A_1^c A_2^c$,

$$P(A_3)=P(A_3A_1^cA_2^c)=P(A_1^c)P(A_2^c\,|\,A_1^c)P(A_3\,|\,A_1^cA_2^c)$$
$$=\left(1-\frac{1}{5}\right)\times\left(1-\frac{1}{4}\right)\times\frac{1}{3}=\frac{1}{5}.$$

In fact, $P(A_4)=P(A_5)=\frac{1}{5}$. We omit the details for saving the length.

We next introduce the other two powerful theorems to deal with complicated events by involving the conditional probabilities.

Definition 1.7.2 Partition of the Sample Space

A sequence of events $\{A_i, i=1,2,\cdots,n\}$, is called a partition of the sample space S if

(1) for any $i,j\in\{1,2,\cdots,n\}$, $i\neq j$, $A_iA_j=\varnothing$,

(2) $\bigcup_{i=1}^{n} A_i = S$.

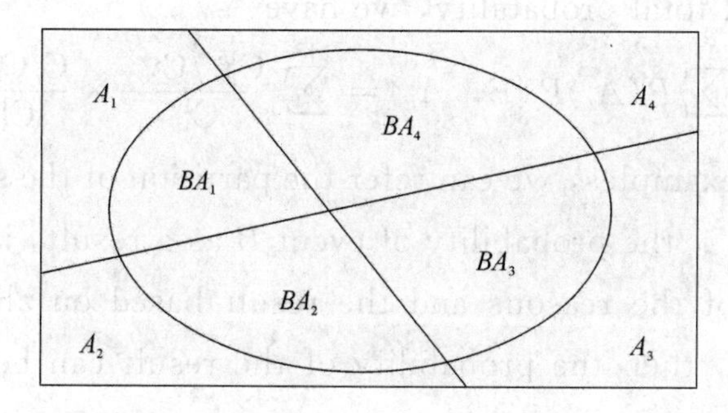

Figure 1.7.1

Theorem 1.7.2 Law of Total Probability

There is a partition $\{A_i, i=1,2,\cdots,n\}$ of the sample space S. $P(A_i)>0$ for all $i=1,2,\cdots,n$. Then for any event B, we have

$$P(B)=\sum_{i=1}^{n}P(A_i)P(B\mid A_i). \tag{1.7.3}$$

Proof.

Since $\{A_i, i=1,2,\cdots,n\}$ is a partition, $B=\bigcup_{i=1}^{n} BA_i$ and $BA_i, i=1,2,\cdots,n$ are mutually exclusive. By the multiplication rule, we have

$$P(B)=\sum_{i=1}^{n}P(BA_i)=\sum_{i=1}^{n}P(A_i)P(B\mid A_i).$$

Example 1.7.5 Continue Example 1.5.5

We choose 10 balls again for a second match, what is the probability p that exactly 3 new balls will be selected.

Solution.

We need a partition of the status of 45 balls after the first match.

A_i, $i=5,6,\cdots,15$: There are i new balls after the first match before the second one.

B: Exactly 3 new balls will be selected for the second match.

$\{A_i, i=5,6,\cdots,15\}$ is a partition, and $P(A_i)=\dfrac{C_{15}^{15-i}C_{30}^{i-5}}{C_{45}^{10}}$.

Further, $P(B|A_i)=\dfrac{C_i^3 C_{45-i}^7}{C_{45}^{10}}$.

By the law of total probability, we have

$$p=\sum_{i=5}^{15}P(A_i)P(B\mid A_i)=\sum_{i=5}^{15}\frac{C_{15}^{15-i}C_{30}^{i-5}}{C_{45}^{10}}\times\frac{C_i^3 C_{45-i}^7}{C_{45}^{10}}.$$

From above examples, we can refer the partition of the sample space as a series of "reasons", the probability of event B as a result, in some sense. If the probabilities of the reasons and the result based on all reasons can be simulated at first, then the probability of the result can be concluded from Theorem 1.7.2.

On the same circumstance, if the result occurred and was observed, how can we readjust the probabilities of the reasons with the updated information from the result?

For example, in Example 1.7.5, result B was observed, what is $P(A_i|B)$? That is the case where Bayes's rule applies well.

Theorem 1.7.3 Bayes's Rule

There is a partition $\{A_i, i=1,2,\cdots,n\}$ of the sample space S. $P(A_i)>0$ for all $i=1,2,\cdots,n$. Then for any event B with $P(B)>0$, we have

$$P(A_i\mid B)=\frac{P(A_i)P(B\mid A_i)}{\sum_{k=1}^{n}P(A_k)P(B\mid A_k)}. \qquad (1.7.4)$$

Proof.

This rule is also an application of the definition of conditional probability with slight extension.

$$P(A_i \mid B) = \frac{P(A_i B)}{P(B)} = \frac{P(A_i)P(B \mid A_i)}{\sum_{k=1}^{n} P(A_k)P(B \mid A_k)}.$$

Bayes's rule gives us a method to adjust our decision by sampling.

Example 1.7.6 Diagnosis of Liver Cancer

Suspected patient of liver cancer is often diagnosed through testing alpha fetoprotein. We denote by A_1 that the patient is suffering from liver cancer, by A_2 that the patient is not. Event B represents that the testing result is positive. Assume by statistics that $P(B|A_1)=0.99$, and $P(B|A_2)=0.05$. According to a survey of some local residents, $P(A_1)=0.0004$. Now, if the positive reaction was tested for a suspected patient, what is the probability that he is suffering from liver cancer?

Solution.

It is easy to see that $A_1 = A_2^c$ and $P(A_2)=0.9996$. The probability that he is suffering from liver cancer is $P(A_1|B)$.

$$P(A_1|B) = \frac{P(A_1)P(B|A_1)}{P(A_1)P(B|A_1)+P(A_2)P(B|A_2)}$$

$$= \frac{0.0004 \times 0.99}{0.0004 \times 0.99 + 0.9996 \times 0.05} = 0.00786.$$

People will wonder the validity of the alpha fetoprotein test since that the probability that he is suffering from liver cancer is so small although the positive reaction is observed. Yet the reason of such a deviation is not difficult to find. That is the small probability that a person is suffering from liver cancer. Actually, doctors use to perform some simple preliminary investigation to exclude a lot of people who do obviously not get liver cancer. When doctors suspect someone who may be suffering from liver cancer, the alpha fetoprotein test is recommended. Then the accuracy will be significantly improved. For example, if $P(A_1)=0.4$, then

$$P(A_1|B)=\frac{P(A_1)P(B|A_1)}{P(A_1)P(B|A_1)+P(A_2)P(B|A_2)}$$

$$=\frac{0.4\times 0.99}{0.4\times 0.99+0.6\times 0.05}=0.9296.$$

In Bayes's rule, a probability like $P(A_i)$ mentioned above is often called the **prior probability**. A probability like $P(A_i|B)$ is then called the **posterior probability**.

1.8 Independent Events

Previously, we measure the chance of the event with updating information. Which one is larger between $P(A|B)$ and $P(A)$? Nothing is impossible here. Let's see an example. A balanced dice is rolled. Suppose that A is that the number is 1 and B is that the number is odd. Obviously, $P(A|B)>P(A)$. Does it mean that the probability will increase with more information known? The answer is NO. For example, $P(A^c|B)<P(A^c)$. Sometimes, the probability will keep unchangeable with more information known. For example, $P(A|S)=P(A)$, which is the independence we talk about in this section: some event B has occurred, which does not change the probability of the other event A, in a word, A and B are independent.

In general, we can define A and B are independent as follows: $P(A|B)=P(A)$, if $P(B)>0$.

However, this definition excludes the events with zero probability. Therefore we take another form.

Definition 1.8.1 Independent

Two events A and B are independent if $P(AB)=P(A)P(B)$.

Example 1.8.1 Roll a Die

Suppose that a balanced die was rolled twice independently. Find the probability p that the first number is 1 and the second number is 2.

Solution.

Let A be the event that the first number is 1 and B be the event that the second number is 2. A and B are independent.

$$P(A)=P(B)=\frac{1}{6}.$$

Then $p=P(AB)=P(A)P(B)=\frac{1}{36}$.

Example 1.8.2

There is a paragraph of encrypted telegram to be deciphered independently by Luren Jia and Luren Yi. Suppose that the probability that Luren Jia will decipher the paragraph is 0.4, Luren Yi 0.5. We shall determine the probability p that this paragraph of encrypted telegram will be deciphered.

Solution.

Let A be the event that Luren Jia will decipher the paragraph. $P(A)=0.4$.

Let B be the event that Luren Yi will decipher the paragraph. $P(B)=0.5$.

Then the event that this paragraph of encrypted telegram will be deciphered is $A\cup B$.

$$\begin{aligned}p&=P(A\cup B)=P(A)+P(B)-P(AB)\\&=P(A)+P(B)-P(A)P(B)=0.4+0.5-0.2=0.7.\end{aligned}$$

Theorem 1.8.1

Suppose that event A and event B are independent. Then

(1) $P(AB^c)=P(A)P(B^c)$.

(2) $P(A^cB^c)=P(A^c)P(B^c)$.

Proof.

For (1),

$$\begin{aligned}P(AB^c)&=P(A-AB)=P(A)-P(AB)=P(A)-P(A)P(B)\\&=P(A)[1-P(B)]=P(A)P(B^c).\end{aligned}$$

For (2),

$$\begin{aligned}P(A^cB^c)&=1-P(A\cup B)=1-P(A)-P(B)+P(AB)\\&=1-P(A)-P(B)+P(A)P(B)=1-P(A)-P(B)[1-P(A)]\\&=[1-P(A)][1-P(B)]=P(A^c)P(B^c).\end{aligned}$$

From the above content, by the independence of event A and event B, we mean that event A will not interfere the chance that event B occurs or not. From the view point of set theory, what is the relationship between A and B?

Example 1.8.3

Suppose that event A and event B satisfy $0<P(A)<1$, $0<P(B)<1$. Which of the following 3 cases possibly implies event A and event B are independent? (1) $AB=\varnothing$, (2) $AB\neq\varnothing$, (3) $A\subset B$.

Solution.

Event A and event B are not independent in (1) and (3), possibly independent in (2).

In (1), $P(AB)=0$, but $P(A)P(B)\neq 0$.

In (3), $P(AB)=P(A)$, but $P(A)P(B)<P(A)$.

Intuitively, case (1) means that event B will definitely not occur if event A occurs. Case (3) means that event B will definitely occur if event A occurs. Both cases should imply event A and event B are not independent. However, why does case (2) possibly imply event A and event B are independent? It seems contradictory to our intuition.

Let's see another example.

Example 1.8.4 Rolling a Die

Suppose that a balanced die is rolled. Let A be the event that an odd number is obtained, and let B be the event that a number less than 5 is obtained. We shall show that the events A and B are independent although they have a nonempty intersection.

Proof.

$A=\{1,3,5\}$ and $B=\{1,2,3,4\}$, $AB=\{1,3\}$.

$P(A)=\frac{1}{2}$, $P(B)=\frac{2}{3}$ and $P(AB)=\frac{1}{3}$.

Therefore,

$$P(AB)=P(A)P(B),$$

which implies the events A and B are independent.

Definition 1.8.2　Independent of Several Events

The n events A_1, A_2, $\cdots$, A_n are independent if for every subset $\{i_1, i_2, \cdots, i_k\} \subset \{1, 2, \cdots, n\}$, these events $A_{i_1}, A_{i_2}, \cdots, A_{i_k}$ satisfy

$$P(A_{i_1} A_{i_2} \cdots A_{i_k}) = P(A_{i_1}) P(A_{i_2}) \cdots P(A_{i_k}).$$

For three events A, B and C, in particular, if they are independent, the following four relations must be satisfied:

(1) $P(AB) = P(A)P(B)$,

(2) $P(AC) = P(A)P(C)$,

(3) $P(BC) = P(B)P(C)$,

(4) $P(ABC) = P(A)P(B)P(C)$.

If (1)-(3) hold except (4), events A, B and C are **pairwise independent.** Some may wonder if the relation in (4) is redundant with (1)-(3) holding. Let's see an example.

Example 1.8.5　Rolling Two Dice

Two balanced dice are rolled. See Example 1.5.1 for the sample space. Let A be the event that the number of the first die is odd. Let B be the event that the number of the second die is even. Let C be the event that both numbers are odd or even.

$$A = \left\{\begin{matrix} (1,1),(1,2),(1,3),(1,4),(1,5),(1,6), \\ (3,1),(3,2),(3,3),(3,4),(3,5),(3,6), \\ (5,1),(5,2),(5,3),(5,4),(5,5),(5,6) \end{matrix}\right\},$$

$$B = \left\{\begin{matrix} (1,2),(2,2),(3,2),(4,2),(5,2),(6,2), \\ (1,4),(2,4),(3,4),(4,4),(5,4),(6,4), \\ (1,6),(2,6),(3,6),(4,6),(5,6),(6,6) \end{matrix}\right\},$$

$$C = \left\{\begin{matrix} (1,1),(1,3),(1,5),(3,1),(3,3),(3,5), \\ (5,1),(5,3),(5,5),(2,2),(2,4),(2,6), \\ (4,2),(4,4),(4,6),(6,2),(6,4),(6,6) \end{matrix}\right\}.$$

$AB = \{(1,2),(3,2),(5,2),(1,4),(3,4),(5,4),(1,6),(3,6),(5,6)\}$,

$BC = \{(2,2),(4,2),(6,2),(2,4),(4,4),(6,4),(2,6),(4,6),(6,6)\}$,

$CA = \{(1,1),(1,3),(1,5),(3,1),(3,3),(3,5),(5,1),(5,3),(5,5)\}$.

$ABC=\varnothing$.

Then we can see that

$$P(A)=P(B)=P(C)=\frac{1}{2}, P(AB)=P(BC)=P(CA)=\frac{1}{4},$$

$$P(AB)=P(A)P(B), P(BC)=P(B)P(C), P(CA)=P(C)P(A).$$

$$P(ABC)=0\neq P(A)P(B)P(C).$$

Intuitively, if AB occurs, which means the first number is odd and the second number is even, then C cannot happen. They are not independent.

Exercises:

1.1 Roll 3 dice to observe the sum of the three numbers. Determine the sample space.

1.2 There are three events A, B and C. Determine the following events with A, B and C and their complements.

(1) Only A occurs. (2) A and B occur, but C does not. (3) At least two of three events A, B and C occur. (4) At most one event of three events A, B and C occurs.

1.3 (1) Suppose that $A\subset B$. Show that $B^c\subset A^c$.

(2) For every two events A, B, show that $A=AB\cup AB^c$.

(3) For n events $A_1, A_2, \cdots, A_n$, show that $\left(\bigcup_{i=1}^{n} A_i\right)^c=\bigcap_{i=1}^{n} A_i^c$.

1.4 Suppose that one card is to be selected from a deck of ten cards that contains 5 red cards numbered from 1 to 5 and 5 blue cards numbered from 1 to 5. Let A be the event that a card with an even number is selected; let B be the event that a blue card is selected; and let C be the event that a card with a number less than 3 is selected. Describe the sample space S and describe each of the following events both in words and sets:

(1) ABC, (2) $A\cup B\cup C$, (3) $A(B\cup C)$, (4) $A^c(B\cup C)$.

1.5 The human blood-type system consists of four blood types: A, B, AB, and O. There are two antigens, anti-A and anti-B, which react with a person's blood in different ways depending on the blood type. Antigen anti-A reacts with blood types A and AB, but not with B and O. Antigen anti-B reacts with blood types B and AB, but not with A and O. Suppose that a person's blood is tested with the two antigens. Let E_1 be the event that the blood reacts with anti-A, and let E_2 be the event that it reacts with anti-B. Classify the person's blood type using the events E_1 and E_2

1.6 There are a black balls and b white balls in a bag. $k(k\leqslant a+b)$ balls are taken out one by one without replacement. Compute the probability that the kth ball taken out is black.

1.7 There are 100 students in one class. Compute the probability that at least two students have the same birthday.

1.8 Among a group of 150 students, 70 students are enrolled in a mathematics class, 80 students are enrolled in an English class, and 90 students are enrolled in a music class. Furthermore, the number of students enrolled in both the mathematics and English classes is 40; the number enrolled in both the English and music classes is 50; and the number enrolled in both the mathematics and music classes is 40. Finally, the number of students enrolled in all three classes is 20. We shall determine the probability that a student selected at random from the group of students will be enrolled in none of the three classes.

1.9 There are two events A and B with $P(B)>0$. Prove that $P(A^c|B)=1-P(A|B)$.

1.10 If A and B are two events such that $A\subset B, P(A)=0.4$, $P(B)=0.6$, compute $P(B|A^c)$.

1.11 (Polya Urn) There are n black balls and m white balls in a bag. Suppose that one ball is selected at random, and then put it back with another k balls of the same color. Repeat above steps 2 times. We finally take one ball at the second time. Determine the probability that it is black.

1.12 Consider a machine that produces a defective item with probability p and produces a non-defective item with probability $q=1-p$.

Suppose that items produced by the machine are selected at random and inspected one at a time until exactly 5 defective items have been obtained. We shall determine the probability p_n that exactly n items ($n>5$) must be selected to obtain the 5 defective items.

1.13 If A, B and C are independent events such that $P(A)=1/2$, $P(B)=1/3$, $P(C)=1/4$, compute $P(A\cup B\cup C)$.

1.14 Two boxes contain long bolts and short bolts. Suppose that one box contains 40 long bolts and 20 short bolts, and that the other box contains 10 long bolts and 20 short bolts. Suppose also that one box is selected at random and a bolt is then selected at random from that box. Determine the probability that this bolt is long.

1.15 A large batch of similar items are produced by three machines M_1, M_2 and M_3. Suppose that 10 percent of the items were produced by machine M_1, 30 percent by machine M_2, and 60 percent by machine M_3. Suppose further that 5 percent of the items produced by machine M_1 are defective, that 3 percent of the items produced by machine M_2 are defective, and that 1 percent of the items produced by machine M_3 are defective. (1) Suppose that one item is selected at random from the entire batch. Determine the probability that it is found to be defective. (2) Suppose that one item is selected at random from the entire batch and found to be defective. We shall determine the probability that this item was produced by machine M_2.

1.16* (The Monty Hall problem) Suppose you're on a game show, and you're given the choice of three doors: Behind one door is a car; behind the others, goats. Of course, you don't know what is behind the door. Once you open a door, you will get what is behind the door. Now you pick a door, say No. 1, and the host of the game, who knows what's behind the doors, opens another door, say No. 3, behind which is a goat. He then says to you, "Do you want to pick door No. 2?" Is it more possible to get the car to switch your choice?

1.17* (Optimal gambler's strategy) A gambler has one hundred dollars at the beginning. Each time, a balanced coin is tossed to decide who is the winner. The host bets the same amount of money as the gambler. The winner claims all the wagers. Suppose that the gambler will leave the game immediately once he wins an expected amount of money or lose everything. By Doob's inequality, the maximum probability, of which the gambler leaves with 200 dollars, is $\frac{1}{2}$. (1) Please construct a strategy for the gambler such that he achieves the maximum probability. (2) Further, still from Doob's inequality, the maximum probability, that the gambler leaves with 300 dollars, is $\frac{1}{3}$. Please construct a strategy for the gambler such that he achieves the maximum probability.

The exercises with mark * are proposed to those who are interested.

Chapter 2 Random Variable and Distribution

2.1 Random Variable

We change our method to represent an event by involving functions. A (measurable) function constructed maps sample space to real space **R**. **R** is always the real space through this book without special explanation. When the function takes value in some real set (measurable set), the inverse of the function is some subset of the sample space, i. e. , some specific event. Then we change our way of measuring an event into one of measuring how the function takes values, which brings us great convenience and improves the probability theory. The function is the random variable. Here we do not take time on explaining the notation of "measurable". In fact, it is a little complicated for the second year student.

Let's see some examples and then form the definition of a random variable.

Example 2.1.1 Tossing a Coin

Consider an experiment in which a coin is tossed 3 times. In this experiment the sample space consists of 2^3 outcomes:

$$S=\{HHH, HHT, HTH, THH, HTT, THT, TTH, TTT\},$$

where H means head, and T means tail.

Solution.

Let the function X be the number of heads obtained on the 3 tosses. Then it is easy to see that

$$X: S \to \{0,1,2,3\} \subset \mathbf{R}.$$

If sample $s_1 = \mathrm{HTH}$, $X(s_1)=2$. If sample $s_2 = \mathrm{HTT}$, $X(s_2)=1$.

Let A be an event that we have 2 heads in the experiment. Then

$$A = \{s \in S, X(s)=2\} = \{\mathrm{HHT}, \mathrm{HTH}, \mathrm{THH}\}.$$

Let B be an event that we have at most 2 heads in the experiment. Then

$$B = \{s \in S, X(s) \leqslant 2\} = \{\mathrm{HHT}, \mathrm{HTH}, \mathrm{THH}, \mathrm{HTT}, \mathrm{THT}, \mathrm{TTH}, \mathrm{TTT}\}.$$

Of course, one function is far from enough to represent all events. For example, let C be an event that we have a head in the first toss. It is impossible to represent C by X no matter how X takes values. However, we can propose another function $Y: S \to \{0,1\}$, where Y is 0 if the tail is observed in the first toss and 1 otherwise. Now we are ready for C.

$$C = \{s \in S: Y(s)=1\} = \{\mathrm{HHH}, \mathrm{HHT}, \mathrm{HTH}, \mathrm{HTT}\}.$$

Example 2.1.2 Selecting Numbers at Random.

Consider an experiment in which a number is selected from an interval $[0,1]$ at random. Let the function X be the number selected. The sample space $S=[0,1]$.

Solution.

$X: [0,1] \to [0,1] \subset \mathbf{R}$. $X(s)=s$, which is an identical mapping. The event A that the value is greater than 0.6 can be represented by

$$A = \{s \in S: X(s) > 0.6\}.$$

Suppose all measurable events are contained in some set $\mathscr{F}$ (σ algebra) and the probability measure is denoted by P. We call the triplet $(S, \mathscr{F}, P)$ the **probability space**. Here $\mathscr{F}$ is a collection of sets.

Definition 2.1.1 Random Variable

Suppose that S is the sample space in an experiment and the probability space is $(S, \mathscr{F}, P)$. A real-valued function X that is defined on the space S is called a random variable if for any

$$A=\{s\in S: X(s)\in(-\infty,x],x\in\mathbf{R}\}\in\mathscr{F},$$

that is, A is measurable.

In the following, we denote $\{s\in S: X(s)\in B\}$ by $\{X\in B\}$ for short.

Remark 2.1.1

The random variable is not necessary to be a one-dimensional function. This case will be explained in the next chapter.

Example 2.1.3 Tossing a Coin (continue Example 2.1.1)

Find the probability p_1 that we have at least one head, and the probability p_2 that we have tail in the first toss.

Solution.

$$p_1=P\{X\geqslant 1\}=\frac{7}{8},\quad p_2=P\{Y=0\}=\frac{4}{8}=\frac{1}{2}.$$

Since we do not go deeply into the measure theory, only two special kinds of random variables are considered: discrete type and continuous type. By the **discrete random variable**, we mean that the random variable takes only countable number of values. As to the **continuous random variable**, it takes uncountably infinite values which fill a whole interval. Sometimes we denote random variable by "r. v." for short.

Next we turn to measure the event represented by the random variable: $P\{X\in B\}$ by using distribution function. No matter if the r. v. is discrete or not, we give the definition of distribution function (d. f.).

Definition 2.1.2 Distribution Function

The distribution function F of a random variable X is a function defined as follows:

$$F(x)=P\{X\leqslant x\}\text{ for each real number }x.$$

Here the event of this form $\{X\in(-\infty,x]\}$ is one measurable set in the σ algebra.

The expression $X\sim F(x)$ means that the distribution function of r. v. X is $F(x)$. $F(x)$ is the probability of the event that the r. v. X takes value from $(-\infty,x]$.

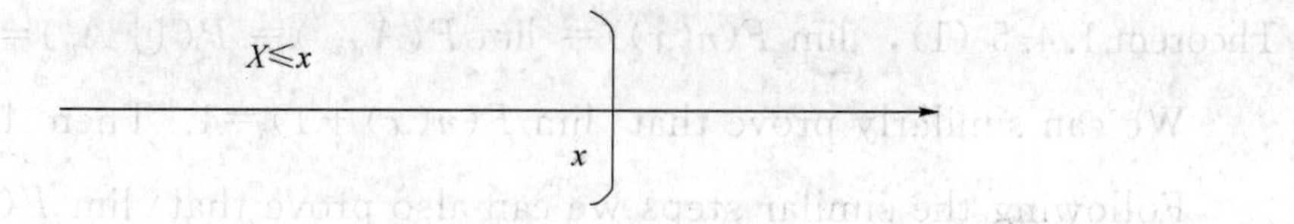

Figure 2.1.1

In fact, $F(x)$ is used to calculate the probabilities of all kinds of events represented by X. Before we do this, let's see the following properties for the distribution function.

Theorem 2.1.1

Suppose that $X \sim F(x)$.

(1) $0 \leqslant F(x) \leqslant 1$ for $x \in \mathbf{R}$;

(2) $F(x)$ is a non-decreasing function;

(3) $\lim\limits_{x \to +\infty} F(x)=1$, $\lim\limits_{x \to -\infty} F(x)=0$;

(4) $F(x)$ is right continuous.

Remark 2.1.2

A function $f(x)$ is right continuous at some point x_0, if $\lim\limits_{n \to +\infty} f(x_n)= f(x_0)$ for any sequence $\{x_n\}_{n \geqslant 1}$ with $x_n \geqslant x_0$ and $\lim\limits_{n \to +\infty} x_n = x_0$. For such a sequence $\{x_n\}_{n \geqslant 1}$, we denote the symbol $\lim\limits_{n \to +\infty} x_n = x_0$ by $x_n \to x_{0+}$. That is, x_n tends to x_0 from the right side. Similarly, $x_n \to x_{0-}$ means that x_n tends to x_0 from the left side.

$f(x_{0+})$ ($f(x_{0-})$) denotes the right (left) limit of $f(x)$ at x_0.

Proof.

(1) is obvious.

(2) For any $x_1, x_2 \in \mathbf{R}, x_1 \leqslant x_2$,

$$\{s \in S: X(s) \leqslant x_1\} \subset \{s \in S: X(s) \leqslant x_2\}.$$

Hence, $F(x_1) \leqslant F(x_2)$.

(3) Define $n(x)=\lfloor x \rfloor$, which is the largest integer no greater than x. $F(n(x)) \leqslant F(x) \leqslant F(n(x)+1)$ from (2). For fixed x,

Define $A_k = \{X \leqslant k\}, k=1,2,\cdots,n(x)$, which are increasing. Then from

Theorem 1.4.5 (1), $\lim\limits_{x\to+\infty} F(n(x)) = \lim\limits_{x\to+\infty} P(A_{n(x)}) = P(\bigcup\limits_{n=1}^{\infty} A_n) = P(S) = 1$.

We can similarly prove that $\lim\limits_{x\to+\infty} F(n(x)+1)=1$. Then $\lim\limits_{x\to+\infty} F(x)=1$.

Following the similar steps we can also prove that $\lim\limits_{x\to-\infty} F(x)=0$ by the continuity of probability. The details are omitted here.

(4) To prove the right continuous, it suffices to prove that

$\lim\limits_{n\to\infty} F(x_n) = F(x_0)$ for any $x_0 \in \mathbf{R}$, and any decreasing sequence of $\{x_n\}_{n\geqslant 1}$ with $x_n > x_0$ and $\lim\limits_{n\to-\infty} x_n = x_0$.

Define $A_n = \{X \leqslant x_n\}$. Then $A_1 \supset A_2 \supset \cdots \supset A_n \supset \cdots$ and $A = \bigcap\limits_{n=1}^{\infty} A_n = \{X \leqslant x_0\}$. From Theorem 1.4.5 (2),

$$\lim_{n\to\infty} F(x_n) = \lim_{n\to\infty} P(A_n) = P(\bigcap_{n=1}^{\infty} A_n) = P\{X \leqslant x_0\} = F(x_0).$$

Example 2.1.4

Is $F(x)$ defined below a d. f. of some r. v. ?

$$F(x) = \begin{cases} \sin x, & 0 \leqslant x \leqslant \pi; \\ 0, & \text{otherwise}. \end{cases}$$

Solution.

It is not a d. f. because it is not non-decreasing function.

Example 2.1.5

Suppose that $X \sim F(x)$, where $F(x) = \begin{cases} A + Be^{-\lambda x}, & x > 0; \\ 0, & x \leqslant 0. \end{cases}$ λ is a positive parameter. Find A and B.

Solution.

Since $\lim\limits_{x\to+\infty} F(x) = 1$, $A = 1$. From Theorem 2.1.1 (4) and the expression of $F(x)$,

$$0 = F(0) = \lim_{x\to 0_+} F(x) = \lim_{x\to 0_+} Be^{-\lambda x} + 1 \Rightarrow B = -1.$$

Now we consider how $F(x)$ describes the probabilities of all kinds of events.

Theorem 2.1.2

Suppose that $X \sim F(x)$.

(1) $P\{a<X\leqslant b\}=F(b)-F(a)$.

(2) $P\{X>b\}=1-F(b)$.

(3) $P\{a<X<b\}=F(b_-)-F(a)$, where $F(b_-)$ is the left limit of $F(x)$ at b.

(4) $P\{X<b\}=F(b_-)$.

(5) $P\{X=b\}=F(b)-F(b_-)$.

(6) $P\{X\geqslant b\}=1-F(b_-)$.

Proof.

(1) $P\{a<X\leqslant b\}=P\{\{X\leqslant b\}-\{X\leqslant a\}\}=P\{X\leqslant b\}-P\{X\leqslant a\}=F(b)-F(a)$.

(2) $P\{X>b\}=P\{\{X\leqslant b\}^c\}=1-F(b)$.

(3) Define a sequence of events $A_n=\left\{a<X\leqslant b-\frac{1}{n}\right\}, n=1,2,\cdots$.

It is easy to see that $A_1\subset A_2\subset\cdots\subset A_n\subset\cdots$ and $\bigcup\limits_{n=1}^{\infty} A_n=\{a<X<b\}$.

Then $P\{a<X<b\}=\lim\limits_{n\to\infty}P(A_n)=\lim\limits_{n\to\infty}F\left(b-\frac{1}{n}\right)-F(a)=F(b_-)-F(a)$.

(4) By slightly changing the proof of (3) with $A_n=\left\{X\leqslant b-\frac{1}{n}\right\}, n=1,2,\cdots$, we have $P\{X<b\}=F(b_-)$ immediately.

(5) $P\{X=b\}=P\{\{X\leqslant b\}-\{X<b\}\}=P\{X\leqslant b\}-P\{X<b\}=F(b)-F(b_-)$.

(6) $P\{X\geqslant b\}=1-P\{X<b\}=1-F(b_-)$.

Example 2.1.6 Continue Example 2.1.1

Determine the distribution function of X.

Solution.

$F(x)=P\{X\leqslant x\}$. It is known that the range of X is $\{0,1,2,3\}$.

For $x\geqslant 3$, $F(x)=P\{X\leqslant x\}=P(S)=1$.

For $2\leqslant x<3$, $F(x)=P\{X\leqslant x\}=P\{X\in\{0,1,2\}\}$

$$=\mathrm{P}\{TTT,HTT,THT,TTH,HHT,HTH,THH\}$$

$$=\frac{7}{8}.$$

For $1\leqslant x<2$, $F(x)=P\{X\leqslant x\}=P\{X\in\{0,1\}\}=\frac{1}{2}$.

For $0\leqslant x<1$, $F(x)=P\{X\leqslant x\}=P\{X=0\}=\frac{1}{8}$.

For $x<0$, $F(x)=P\{X\leqslant x\}=P(\phi)=0$.

In summary, we have

$$F(x)=\begin{cases}0, & x<0;\\ \frac{1}{8}, & 0\leqslant x<1;\\ \frac{1}{2}, & 1\leqslant x<2;\\ \frac{7}{8}, & 2\leqslant x<3;\\ 1, & x\geqslant 3.\end{cases}$$

2.2 Discrete Distribution

In this section, we focus on the discrete random variable and its distribution. In view of the discrete values, a new function is introduced to help measure the corresponding events.

Definition 2.2.1 Probability Function (p. f.)

If a discrete r. v. X can take only finite values $x_1, x_2, \cdots, x_n$ (or, at most, a sequence of infinite values $x_1, x_2, \cdots, x_n, \cdots$), define $P\{X=x_k\}=p_k$, for $k=1, 2, \cdots, n$ (or $k=1, 2, \cdots, n, \cdots$). $\{p_k, k=1,2,\cdots,n\}$ (or $\{p_k, k=1,2,\cdots,n,\cdots\}$) is called the probability function (p. f.) of r. v. X.

Obviously, the p. f. should satisfy the following properties:

Theorem 2.2.1

For a p. f. $\{p_k, k=1,2,\cdots,n\}$ (or $\{p_k, k=1,2,\cdots\}$),

(1) $p_k\geqslant 0, k=1,2,\cdots,n$ (or $p_k\geqslant 0, k=1,2,\cdots$).

(2) $\sum_{k=1}^{n} p_k=1$ (or $\sum_{k=1}^{\infty} p_k=1$).

Proof.

We give only the proof that X takes finite values $x_1, x_2, \cdots, x_n$.

(1) is obvious.

For (2), $\{X=x_k\}, k=1,2,\cdots,n$ is a partition of S. Then

$$1 = P(S) = P\{\bigcup_{k=1}^{n} \{X = x_k\}\} = \sum_{k=1}^{n} P\{X = x_k\} = \sum_{k=1}^{n} p_k.$$

Example 2. 2. 1

Suppose that the range of discrete r. v. X is confined to the set of all nonnegative integers. $P\{X=k\}=a\frac{\lambda^k}{k!}, k=0,1,2,\cdots$, where λ is a positive parameter. Determine a.

Solution.

From Theorem 2. 2. 1 (2), $\sum_{k=0}^{\infty} a\frac{\lambda^k}{k!} = 1 \Rightarrow a = e^{-\lambda}$.

There are usually 3 ways to display the probability function, one of which is just as in Example 2. 2. 1, shown by a formula,

$$P\{X=k\}=\frac{\lambda^k}{k!}e^{-\lambda}, k=0,1,2,\cdots.$$

Another one is to use tables as follows:

Table 2. 2. 1

X	0	1	2	$\cdots$	n	$\cdots$
$P\{X=k\}$	$e^{-\lambda}$	$\lambda e^{-\lambda}$	$\frac{\lambda^2}{2!}e^{-\lambda}$	$\cdots$	$\frac{\lambda^n}{n!}e^{-\lambda}$	$\cdots$

The third one is to use figures.

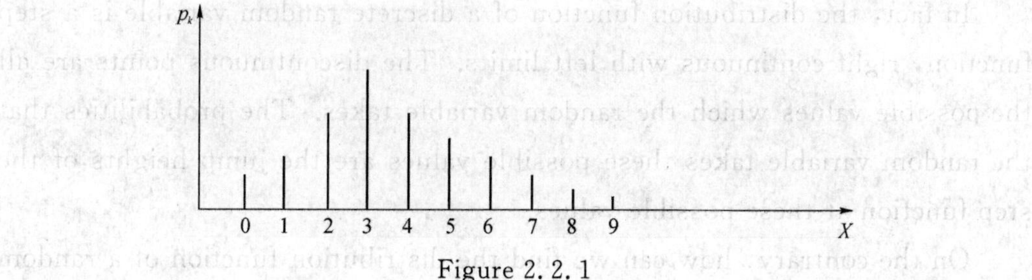

Figure 2. 2. 1

Example 2.2.2 Continue Example 2.1.6

We shall determine the p. f. of X.

Solution.

We find the p. f. by 2 methods.

(1) Direct calculation from the probability model. There are 4 possible values which X takes: $\{0,1,2,3\}$.

$$P\{X=0\}=\frac{1}{8},\quad P\{X=1\}=\frac{3}{8},\quad P\{X=2\}=\frac{3}{8},\quad P\{X=3\}=\frac{1}{8}.$$

(2) Distribution function method. In view of

$$F(x)=\begin{cases}0, & x<0;\\ \frac{1}{8}, & 0\leqslant x<1;\\ \frac{1}{2}, & 1\leqslant x<2;\\ \frac{7}{8}, & 2\leqslant x<3;\\ 1, & x\geqslant 3,\end{cases}$$

we have

$$P\{X=3\}=F(3)-F(3_-)=1-\frac{7}{8}=\frac{1}{8},$$

$$P\{X=2\}=F(2)-F(2_-)=\frac{7}{8}-\frac{1}{2}=\frac{3}{8},$$

$$P\{X=1\}=F(1)-F(1_-)=\frac{1}{2}-\frac{1}{8}=\frac{3}{8},$$

$$P\{X=0\}=F(0)-F(0_-)=\frac{1}{8}-0=\frac{1}{8}.$$

In fact, the distribution function of a discrete random variable is a step function, right continuous with left limits. The discontinuous points are all the possible values which the random variable takes. The probabilities that the random variable takes these possible values are the jump heights of the step function at these possible values.

On the contrary, how can we find the distribution function of a random variable with the probability function known? Let's see the following example.

Example 2.2.3

Suppose that the p. f. of a r. v. is

$$P\{X=0\}=\frac{1}{8},\quad P\{X=1\}=\frac{3}{8},\quad P\{X=2\}=\frac{3}{8},\quad P\{X=3\}=\frac{1}{8}.$$

We shall determine the d. f. $F(x)$.

For $x<0$, $P\{X\leqslant x\}=P(\phi)=0$.

For $x\geqslant 3$, $P\{X\leqslant x\}=P(S)=1$.

For $0\leqslant x<1$, $P\{X\leqslant x\}=P\{X=0\}=\frac{1}{8}$.

For $1\leqslant x<2$, $P\{X\leqslant x\}=P\{X=0\}+P\{X=1\}=\frac{1}{8}+\frac{3}{8}=\frac{1}{2}$.

For $2\leqslant x<3$, $P\{X\leqslant x\}=P\{X=0\}+P\{X=1\}+P\{X=2\}=\frac{1}{8}+\frac{3}{8}+\frac{3}{8}=\frac{7}{8}$.

Then we have the d. f. $F(x)=\begin{cases}0, & x<0;\\ \frac{1}{8}, & 0\leqslant x<1;\\ \frac{1}{2}, & 1\leqslant x<2;\\ \frac{7}{8}, & 2\leqslant x<3;\\ 1, & x\geqslant 3.\end{cases}$

In summary, we can easily find the relationship between the probability function and the distribution function of a random variable in the following theorem.

Theorem 2.2.2

Suppose that the possible values of r. v. X are $x_1, x_2, \cdots, x_k, \cdots$, the p. f. is $p_k=P\{X=x_k\}$, and the d. f. is $F(x)$. Then

(1) $F(x)=\sum_{\{k: x_k\leqslant x\}} p_k$,

(2) $p_k = F(x_k) - F(x_k -)$,

(3) $P\{X \in A\} = \sum_{\{k: x_k \in A\}} p_k$.

The proof is omitted since the theorem is direct.

Example 2. 2. 4

Suppose r. v. X takes values from $\{-1,0,1\}$ and the p. f. of X is $P\{X=1\}=P\{X=-1\}=\frac{1}{4}$, $P\{X=0\}=a$. Determine a and $P\{|X|=1\}$.

Solution.

$$\frac{1}{4}+\frac{1}{4}+a=1 \Rightarrow a=\frac{1}{2}.$$

$$P\{|X|=1\}=P\{X=1\}+P\{X=-1\}=\frac{1}{2}.$$

Next we will introduce some usual p. f. 's.

1. (0-1) distribution

Suppose that for r. v. X, $P\{X=0\}=p$, $P\{X=1\}=1-p$ for some positive parameter $p \in (0,1)$.

Example 2. 2. 5

A biased coin is tossed. Define that r. v. X is 0 when we have a tail, 1 when we have a head. Then r. v. X obeys the (0-1) distribution.

The experiment where there are exact two outcomes as in Example 2. 2. 5 is called a Bernoulli trial.

2. Binomial Distribution

Suppose that for r. v. X, $P\{X=k\}=C_n^k p^k (1-p)^{n-k}$, $k=0,1,2,\cdots,n$, $p \in (0,1)$. Then we say r. v. X obeys the binomial distribution with parameter (n,p), which is denoted by $X \sim B(n,p)$ for short.

Example 2. 2. 6 Continue Example 2. 2. 5

We toss the coin independently n times. Let X be the r. v. , the number of tails we have. We shall determine the p. f. of X.

Solution.

Let A_k be the event that we have the tail in the kth toss.

(1) There are n cases that we have one tail in n tosses:

$A_1A_2^cA_3^c\cdots A_n^c$, $A_1^cA_2A_3^c\cdots A_n^c$, $A_1^cA_2^cA_3\cdots A_n^c$, $\cdots$, $A_1^cA_2^cA_3^c\cdots A_n$

For each case, the probability is $p(1-p)^{n-1}$ from the independence.

Then $P\{X=1\}=np(1-p)^{n-1}=C_n^1p(1-p)^{n-1}$.

(2) There are C_n^2 cases that we have two tails in n tosses:

$A_1A_2A_3^c\cdots A_n^c$, $A_1^cA_2A_3\cdots A_n^c$, $\cdots$, $A_1^cA_2^cA_3^c\cdots A_{j-1}^cA_jA_{j+1}^c\cdots A_{k-1}^cA_kA_{k+1}^c\cdots A_n^c$, $\cdots$, $A_1^cA_2^cA_3^c\cdots A_{n-2}^cA_{n-1}A_n$.

For each case, the probability is $p^2(1-p)^{n-2}$ from the independence.

Then $P\{X=2\}=C_n^2p^2(1-p)^{n-2}$.

(3) With similar analysis, $P\{X=k\}=C_n^kp^k(1-p)^{n-k}$.

Therefore, $X\sim B(n,p)$.

When $n=1$, X turns to obey the (0-1) distribution and the experiment turns to be a Bernoulli trial. Here n times of Bernoulli trials are performed.

Now we have to face another interesting challenge, calculation with high complexity. In this example, if we set $n=100$, $p=0.05$, then $P\{X=10\}=C_{100}^{10}(0.05)^{10}(0.95)^{90}$. Obviously, it takes time to calculate the exact number of this probability.

In history, French scientist, Poisson (1837) developed the following result to deal with the above problem.

Theorem 2.2.3 Poisson Approximation

Suppose that r. v. $X\sim B(n,p_n)$ with $np_n=\lambda$. Then

$$\lim_{n\to\infty}P\{X=k\}=\frac{\lambda^k}{k!}e^{-\lambda}.$$

Proof.

Since r. v. $X\sim B(n,p_n)$,

$$\begin{aligned}\lim_{n\to\infty}P\{X=k\}&=\lim_{n\to\infty}C_n^k(p_n)^k(1-p_n)^{n-k}\\&=\lim_{n\to\infty}\frac{n\times(n-1)\times\cdots\times(n-k+1)}{k!}\left(\frac{\lambda}{n}\right)^k\left(1-\frac{\lambda}{n}\right)^{n-k}\\&=\frac{\lambda^k}{k!}\lim_{n\to\infty}\frac{n\times(n-1)\times\cdots\times(n-k+1)}{n^k}\left(1-\frac{\lambda}{n}\right)^{n-k}\end{aligned}$$

$$=\frac{\lambda^k}{k!}\lim_{n\to\infty}\frac{n\times(n-1)\times\cdots\times(n-k+1)}{n^k}\left(\left(1-\frac{\lambda}{n}\right)^{-\frac{n}{\lambda}}\right)^{-\frac{n-k}{n}\lambda}=\frac{\lambda^k}{k!}e^{-\lambda}.$$

Corollary 2. 2. 1 Poisson Approximation

Suppose that r. v. $X\sim B(n,p)$. From above theorem, for large n and small p,

(1) $P\{X=k\}=C_n^k p^k(1-p)^{n-k}\approx\frac{\lambda^k}{k!}e^{-\lambda}$, where $\lambda=np$.

(2) $P\{X\geqslant k\}=\sum_{i\geqslant k}^{n}C_n^i p^i(1-p)^{n-i}\approx\sum_{i\geqslant k}^{\infty}\frac{\lambda^i}{i!}e^{-\lambda}$.

Example 2. 2. 7 Application of Poisson Approximation

Suppose that there are 300 sets of equipment, each working independently, in some factory. The probability that one set of equipment breaks down is 0. 01. To ensure the equipment is working properly, the right amount of maintenance personnel are needed. Under normal circumstances, if a set of equipment that breaks down can be handled by one person. How many maintenance personnel should be equipped at least in order to ensure that the probability that the equipment is not timely maintained when it breaks down is less than 0. 01?

Solution.

Let X be the r. v. which is the number of the sets of equipment breaking down. Each set works independently with the failure probability 0. 01. Then

$$X\sim B(300,0.01).$$

Assume that the amount of maintenance personnel is N. What we want is the smallest N, such that

$$P\{X>N\}<0.01.$$

That is,

$$\sum_{k=N+1}^{300}C_{300}^k 0.01^k 0.99^{300-k}<0.01.$$

From Poisson approximation,

$$\sum_{k=N+1}^{300}C_{300}^k 0.01^k 0.99^{300-k}\approx\sum_{k=N+1}^{\infty}\frac{3^k}{k!}e^{-3}=1-\sum_{k=1}^{N}\frac{3^k}{k!}e^{-3}.$$

Then $\sum_{k=1}^{N} \frac{3^k}{k!} e^{-3} \geqslant 0.99$, which implies $N \geqslant 8$. Finally, at least 8 persons are needed.

The limit in Poisson approximation is also a p. f., which formulates many stochastic phenomena well.

3. Poisson Distribution

Suppose that for r. v. X, $P\{X=k\}=\frac{\lambda^k}{k!}e^{-\lambda}$, $k=0,1,2,\cdots$, $\lambda>0$. Then we say r. v. X obeys the Poisson distribution with parameter λ, which is denoted by $X \sim P(\lambda)$ for short.

Now we complete this section by another example of Geometrical distribution.

Example 2.2.8 Geometrical Distribution and Memoryless Property

A shooter will not stop shooting at a target until the target is hit. The probability that he hits the target is p in each round. Suppose that all shots are independent. Denote by X the round number he shoot. We shall determine the p. f. of X.

Solution.

We first determine the possible values which X takes.

The shooter fires at least once to get the target. So the set of possible values is N, containing all the positive integers. We denote the event that the target is hit at the nth round by A_n. Then $P(A_n)=p$ and $A_n, n=1,2,\cdots$ are independent.

$$P\{X=1\}=P\{A_1\}=p,\ P\{X=2\}=P(A_1^c A_2)=P(A_1^c)P(A_2)=(1-p)p,$$
$$P\{X=3\}=P(A_1^c A_2^c A_3)=(1-p)^2 p,\ \cdots,$$
$$P\{X=n\}=P(A_1^c A_2^c \cdots A_{n-1}^c A_n)=(1-p)^{n-1} p,\ \cdots.$$

We call that the r. v. is geometrically distributed with the above p. f.

Suppose that the shooter has fired n rounds and failed to hit the target. Let's determine the probability that he still fails to hit the target after another m rounds, i. e., $P\{X>n+m \mid X>n\}$.

$$P\{X>n+m\mid X>n\}=\frac{P\{X>n+m\}}{P\{X>n\}}$$

$$=\frac{\sum\limits_{k=n+m+1}^{\infty}(1-p)^{k-1}p}{\sum\limits_{k=n+1}^{\infty}(1-p)^{k-1}p}=\frac{(1-p)^{n+m}/p}{(1-p)^{n}/p}=(1-p)^{m}.$$

We notice that the probability is independent of n. In fact, it is exactly the probability that he fails to hit the target in m rounds. That is,

$$P\{X>n+m\mid X>n\}=P\{X>m\}.$$

We call this phenomenon memoryless.

2.3 Continuous Random Variable and Its Distribution

In this section, we focus on the continuous random variable and its distribution. Comparing with the discrete r. v., we introduce a probability density function to help measure the corresponding events.

Roughly speaking, random variables that can take every value in an interval are called continuous.

Definition 2.3.1 (Probability Density Function)

Suppose that a r. v. X can take every value in some interval (or the whole real line). The d. f. is $F(x)$. If there exists a non-negative function $f(x)$, such that $F(x)=\int_{-\infty}^{x}f(t)\mathrm{d}t$, we call X a continuous r. v. and $f(x)$ the probability density function (p. d. f.).

Obviously, the d. f. and p. d. f. of a continuous r. v. should satisfy the following properties:

Theorem 2.3.1

For a p. d. f. $f(x)$ corresponding to a d. f. $F(x)$,

(1) $f(x)\geqslant 0, -\infty<x+\infty$.

(2) $\int_{-\infty}^{+\infty}f(x)\mathrm{d}x=1$.

Proof.

(1) is direct from the definition.

For (2), $\int_{-\infty}^{+\infty} f(x)\mathrm{d}x = \lim_{x \to +\infty} F(x) = 1$.

The p. d. f looks like a linear density. The area of between $f(x)$ and the real line is one.

Example 2.3.1

Suppose that the p. d. f of a continuous r. v. X is $f(x)=\begin{cases} ae^{-2x}, & x \geqslant 0; \\ 0, & x<0. \end{cases}$ Determine a.

Solution.

From Theorem 2.3.1 (2), $\int_{-\infty}^{+\infty} f(x)\mathrm{d}x = \int_{0}^{+\infty} ae^{-2x}\mathrm{d}x = 1 \Rightarrow a = 2$.

Theorem 2.3.2

Let X be a continuous r. v. with p. d. f. $f(x)$ and d. f. $F(x)$.

(1) $P\{a < X \leqslant b\} = \int_a^b f(x)\mathrm{d}x$.

(2) For any real number a, $P\{X=a\}=0$.

(3) $F(x)$ is a continuous function.

(4) $P\{a<X\leqslant b\}=P\{a<X<b\}=P\{a\leqslant X<b\}=P\{a\leqslant X\leqslant b\}=F(b)-F(a)$.

Solution.

(1) $P\{a < X \leqslant b\} = F(b)-F(a) = \int_{-\infty}^{b} f(x)\mathrm{d}x - \int_{-\infty}^{a} f(x)\mathrm{d}x = \int_a^b f(x)\mathrm{d}x$.

(2) $\{X=a\}=\bigcap_{n=1}^{\infty}\left\{a-\frac{1}{n}<X\leqslant a+\frac{1}{n}\right\}$, $\left\{a-\frac{1}{n}<X\leqslant a+\frac{1}{n}\right\}$ is a decreasing sequence of sets.

$$P\{X=a\}=P\left\{\bigcap_{n=1}^{\infty}\left\{a-\frac{1}{n}<X\leqslant a+\frac{1}{n}\right\}\right\}=$$

$$\lim_{n\to+\infty} P\left\{a-\frac{1}{n}<X\leqslant a+\frac{1}{n}\right\}=\lim_{n\to+\infty}\int_{a-\frac{1}{n}}^{a+\frac{1}{n}} f(x)\mathrm{d}x=0.$$

(3) is similar to the proof of (2).

(4) is direct from (1) and (2).

Remark 2.3.1

In fact, for any measurable real set A,

$$P\{X \in A\} = \int_{x \in A} f(x)\mathrm{d}x.$$

By the measurable real set A, we mean that A can be represented by some intervals in the form $(a,b]$ with at most countable times of set algebra calculations, such as union, intersection and complement.

Theorem 2.3.3 Relationship between d. f. and p. d. f.

Suppose that $F(x)$ and $f(x)$ are the d. f. and p. d. f. of r. v. X. Then $f(x) = \frac{\mathrm{d}}{\mathrm{d}x}F(x)$ for any differentiable point x of $F(x)$.

This theorem is direct from the definition of $f(x)$. Let's see an example.

Example 2.3.2

Suppose that X is a continuous r. v. with d. f.

$$F(x) = \begin{cases} 0, & x < 0; \\ x, & 0 \leqslant x \leqslant 1; \\ 1, & x > 1. \end{cases}$$

Find the p. d. f. of X.

Solution.

From Theorem 2.3.3,

$$f(x) = \begin{cases} 1, & 0 \leqslant x \leqslant 1; \\ 0, & \text{otherwise}. \end{cases}$$

We take the derivate of $F(x)$ to get $f(x)$. Is everything OK? No, it is not. For $x \in (-\infty, 0) \cup (0,1) \cup (1, +\infty)$, $F(x)$ is differentiable. However, $F(x)$ is not differentiable at $x = 0, 1$. What is $f(x)$ when $x = 0, 1$? As we know, for a continuous r. v. X,

$$P\{X = 0\} = P\{X = 1\} = \int_{x=0} f(x)\mathrm{d}x = \int_{x=1} f(x)\mathrm{d}x = 0,$$

which implies that no matter what $f(x)$ is at $x = 0, 1$, it will not change the probability that $P\{X \in A\}$. It also makes sense that the p. d. f. is

$$f(x)=\begin{cases}1, & 0<x<1;\\ 0, & \text{otherwise.}\end{cases}$$

The p. d. f. of a continuous r. v. is not unique.

Example 2. 3. 3

Suppose that X is a continuous r. v. with d. f. $F(x)=\begin{cases}Ae^{x}, & x<0;\\ B-Ae^{-x}, & x\geqslant 0.\end{cases}$

Find the p. d. f. of X and $P\{|X|\leqslant 1\}$.

Solution.

We first determine A, B. $\lim\limits_{x\to+\infty} F(x)=1 \Rightarrow B=1$. $\lim\limits_{x\to 0_{+}} F(x)=\lim\limits_{x\to 0_{-}} F(x) \Rightarrow$

$B-A=A$. Then $A=\frac{1}{2}$. Then the p. d. f. of X is

$$f(x)=\begin{cases}\frac{1}{2}e^{x}, & x<0;\\ \frac{1}{2}e^{-x}, & x\geqslant 0.\end{cases}$$

Or we can rewrite the p. d. f as $f(x)=\frac{1}{2}e^{-|x|}, x\in\mathbf{R}$.

Finally, $P\{|X|\leqslant 1\}=\int_{|x|\leqslant 1} f(x)\mathrm{d}x=\int_{-1}^{0}\frac{1}{2}e^{x}\mathrm{d}x+\int_{0}^{1}\frac{1}{2}e^{-x}\mathrm{d}x=1-e^{-1}$.

Next we will introduce some usual p. d. f. 's.

2. 3. 1 Uniform Distribution

Suppose that a continuous r. v. X is uniformly distributed if the p. d. f. is in the following form:

$$f(x)=\begin{cases}\frac{1}{b-a}, & x\in(a,b);\\ 0, & \text{otherwise.}\end{cases}$$

We denote this case by $X\sim U(a,b)$ in brief.

Example 2. 3. 4

Find the d. f. of a continuous r. v. X with the above p. d. f.

Solution.

For $x \leqslant a$, $F(x) = \int_{-\infty}^{x} f(t)\mathrm{d}t = 0$;

For $x \geqslant b$, $F(x) = \int_{-\infty}^{x} f(t)\mathrm{d}t = \int_{a}^{b} \frac{1}{b-a}\mathrm{d}t = 1$;

For $a < x < b$, $F(x) = \int_{-\infty}^{x} f(t)\mathrm{d}t = \int_{a}^{x} \frac{1}{b-a}\mathrm{d}t = \frac{x-a}{b-a}$.

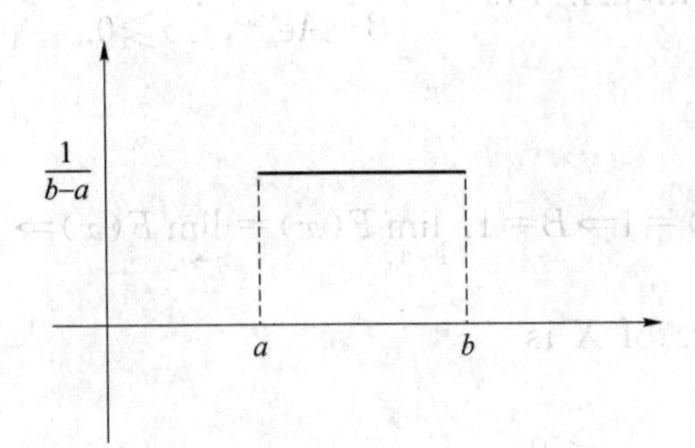

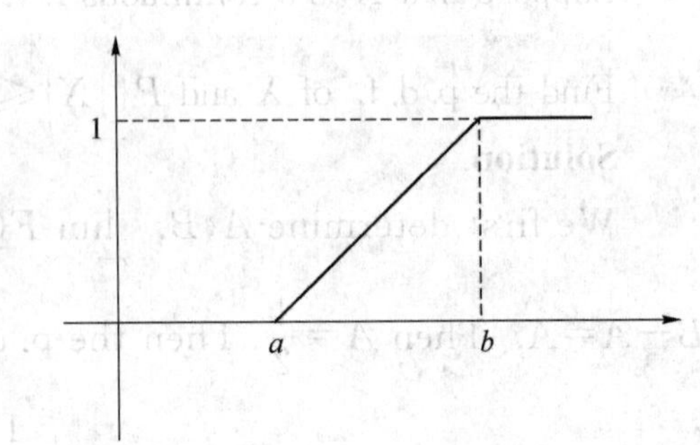

Figure 2. 3. 1

Example 2. 3. 5

A number is selected from an interval (a,b) at random, which is denoted by X. Find the distribution function and construct an event that is possible while the probability is zero.

Solution.

Find the d. f. first.

For $x<a$, $\{X\leqslant x\}$ is impossible. $F(x)=P\{X\leqslant x\}=0$.

For $x>b$, $\{X\leqslant x\}$ is certain. $F(x)=P\{X\leqslant x\}=1$.

For $a\leqslant x\leqslant b$, $\{X\leqslant x\}=\{a\leqslant X\leqslant x\}$. From the geometrical probability model, we have $F(x)=P\{X\leqslant x\}=\frac{x-a}{b-a}$, which is a d. f. of a r. v. with uniform distribution. It is not a surprise since a number is selected at random in this case. The equally likely outcomes mean the density is same everywhere. That is, the r. v. X is uniformly distributed in (a,b). Then we construct the event. Let A be an event that some $c\in(a,b)$ is selected. A is not an empty set, i. e. , A is not impossible. However,

$$P(A)=P\{X=c\}=0.$$

2.3.2 Exponential Distribution

Suppose that a continuous r. v. X is exponentially distributed if the p. d. f. is in the following form:

$$f(x)=\begin{cases}\lambda e^{-\lambda x}, & x\geqslant 0;\\ 0, & \text{otherwise},\end{cases}$$

where λ is a positive parameter.

We denote this case by $X\sim \mathrm{Exp}(\lambda)$ in brief.

Example 2.3.6

Find the d. f. of a continuous r. v. X with the above p. d. f.

Solution.

For $x<0$, $F(x)=\int_{-\infty}^{x} f(t)\mathrm{d}t=0$;

For $x\geqslant 0$, $F(x)=\int_{-\infty}^{x} f(t)\mathrm{d}t=\int_{0}^{x}\lambda e^{-\lambda t}\mathrm{d}t=1-e^{-\lambda x}$.

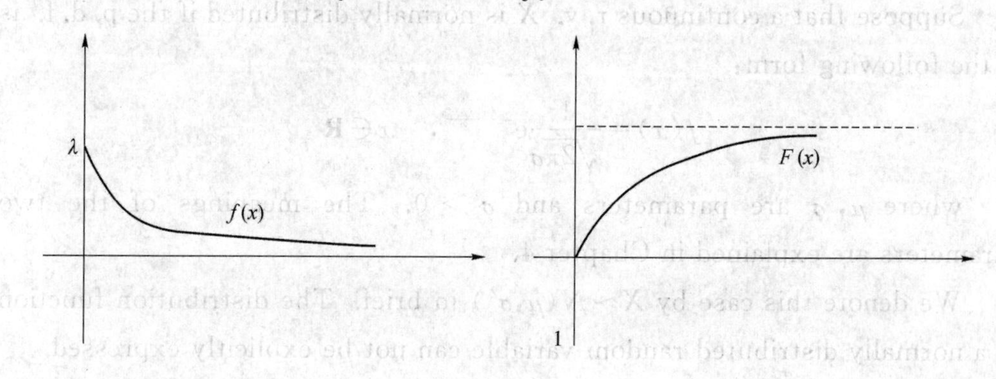

Figure 2.3.2

Example 2.3.7 (Memoryless Property)

Compare with Example 2.2.8. Suppose that the life time (in hours) for some electrical component is exponentially distributed with parameter λ. (1) Find the probability that it can last more than s hours. (2) Suppose that it has lasted for s hours. Determine the probability that it can still last for another t hours.

Solution.

Define r. v. X is the life time for the electrical component. The p. d. f. of X is

$$f(x)=\begin{cases}\lambda e^{-\lambda x}, & x\geqslant 0;\\ 0, & \text{otherwise}.\end{cases}$$

(1) $P\{X>s\}=\int_s^{+\infty}\lambda e^{-\lambda x}\,dx=e^{-\lambda s}$.

(2) $P\{X>s+t\mid X>s\}=\dfrac{P\{X>s+t\}}{P\{X>s\}}=\dfrac{\int_{s+t}^{+\infty}\lambda e^{-\lambda x}\,dx}{\int_s^{+\infty}\lambda e^{-\lambda x}\,dx}=e^{-\lambda t}=$ $P\{X>t\}$.

The memoryless property is also satisfied as in Example 2. 2. 8.

2. 3. 3 Normal Distribution

Suppose that a continuous r. v. X is normally distributed if the p. d. f. is in the following form:

$$f(x)=\frac{1}{\sqrt{2\pi}\sigma}e^{-\frac{(x-\mu)^2}{2\sigma^2}},\quad x\in\mathbf{R}$$

where μ, σ are parameters and $\sigma>0$. The meanings of the two parameters are explained in Chapter 4.

We denote this case by $X\sim N(\mu,\sigma^2)$ in brief. The distribution function of a normally distributed random variable can not be explicitly expressed.

For the special case $\mu=0,\sigma=1$, $N(0,1)$ is called standard normal distribution. We usually denote the d. f. of this case by $\Phi(x)$.

Example 2. 3. 8

Verify $f(x)=\frac{1}{\sqrt{2\pi}\sigma}e^{-\frac{(x-\mu)^2}{2\sigma^2}},x\in\mathbf{R}$ is a p. d. f.

Proof.

It is a non-negative function, obviously.

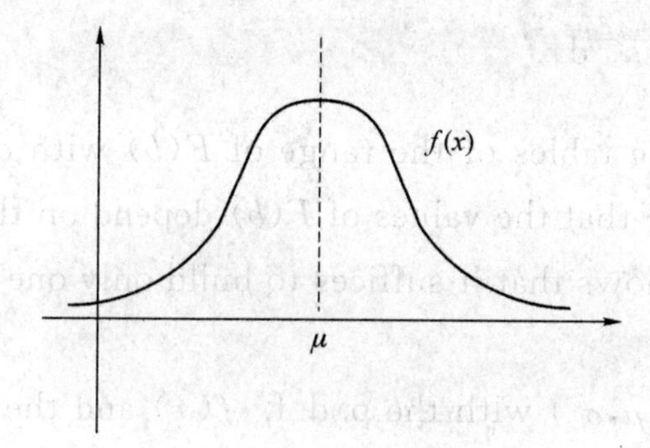

Figure 2. 3. 3

$$\int_{-\infty}^{+\infty} f(x)\mathrm{d}x = \int_{-\infty}^{+\infty} \frac{1}{\sqrt{2\pi}\sigma}\mathrm{e}^{-\frac{(x-\mu)^2}{2\sigma^2}}\mathrm{d}x = \frac{1}{\sqrt{2\pi}}\int_{-\infty}^{+\infty}\mathrm{e}^{-\frac{t^2}{2}}\mathrm{d}t\left(\text{Set } t = \frac{x-\mu}{\sigma}.\right)$$

$$= \left[\frac{1}{\sqrt{2\pi}}\int_{-\infty}^{+\infty}\mathrm{e}^{-\frac{t^2}{2}}\mathrm{d}t \frac{1}{\sqrt{2\pi}}\int_{-\infty}^{+\infty}\mathrm{e}^{-\frac{t^2}{2}}\mathrm{d}t\right]^{\frac{1}{2}} = \frac{1}{\sqrt{2\pi}}\left[\int_{-\infty}^{+\infty}\int_{-\infty}^{+\infty}\mathrm{e}^{-\frac{x^2+y^2}{2}}\mathrm{d}x\mathrm{d}y\right]^{\frac{1}{2}}$$

$$= \frac{1}{\sqrt{2\pi}}\left[\int_0^{2\pi}\int_0^{+\infty}\mathrm{e}^{-\frac{\rho^2}{2}}\rho\,\mathrm{d}\rho\,\mathrm{d}\theta\right]^{\frac{1}{2}} \qquad (\text{Set } x = \rho\cos\theta, y = \rho\sin\theta.)$$

$$= \left[\int_0^{+\infty}\mathrm{e}^{-\frac{\rho^2}{2}}\rho\,\mathrm{d}\rho\right]^{\frac{1}{2}} = \left[-\mathrm{e}^{-\frac{\rho^2}{2}}\Big|_0^{+\infty}\right]^{\frac{1}{2}} = 1.$$

The example is proved.

Let's see the property of normal distribution without proof.

Theorem 2. 3. 4

Suppose $X \sim N(\mu, \sigma^2)$ with the p. d. f. $f(x)$.

(1) $f(\mu - x) = f(\mu + x)$ for any $x \in \mathbf{R}$. That is, $f(x)$ is axisymmetric on a line $x = \mu$.

(2) $f(x)$ achieves the maximum value $\frac{1}{\sqrt{2\pi}\sigma}$ at $x = \mu$.

(3) $\lim_{|x| \to +\infty} f(x) = 0$.

(4) $P\{X < \mu\} = \frac{1}{2}$.

Proof.

It is not easy to find the probability of the event described by a random variable with normal distribution because of the complexity of the integral

$$F(b)=\int_{-\infty}^{b}\frac{1}{\sqrt{2\pi}\sigma}\mathrm{e}^{-\frac{(x-\mu)^2}{2\sigma^2}}\mathrm{d}x.$$

We prepare some tables of the range of $F(b)$ with different values of b for reference. Notice that the values of $F(b)$ depend on the pair (μ,σ^2). The following theorem shows that it suffices to build only one table for reference.

Theorem 2.3.5

Suppose $X\sim N(\mu,\sigma^2)$ with the p. d. f. $f(x)$ and the d. f. $F(x)$.

$$P\{a\leqslant X\leqslant b\}=\Phi\left(\frac{b-\mu}{\sigma}\right)-\Phi\left(\frac{a-\mu}{\sigma}\right)$$

Proof.

$$P\{a\leqslant X\leqslant b\}=\int_{a}^{b}\frac{1}{\sqrt{2\pi}\sigma}\mathrm{e}^{-\frac{(x-\mu)^2}{2\sigma^2}}\mathrm{d}x=\int_{\frac{a-\mu}{\sigma}}^{\frac{b-\mu}{\sigma}}\frac{1}{\sqrt{2\pi}}\mathrm{e}^{-\frac{t^2}{2}}\mathrm{d}t$$

$$=\Phi\left(\frac{b-\mu}{\sigma}\right)-\Phi\left(\frac{a-\mu}{\sigma}\right).$$

Theorem 2.3.6

Suppose $X\sim N(0,1)$. Then

(1) $\Phi(x)+\Phi(-x)=1$.

(2) $\Phi(0)=\frac{1}{2}$.

Proof.

(1) $\Phi(-x)=\frac{1}{\sqrt{2\pi}}\int_{-\infty}^{-x}\mathrm{e}^{-\frac{t^2}{2}}\mathrm{d}t=\frac{1}{\sqrt{2\pi}}\int_{x}^{+\infty}\mathrm{e}^{-\frac{s^2}{2}}\mathrm{d}s=1-\Phi(x)$.

(2) is direct from Theorem 2.3.4 (4).

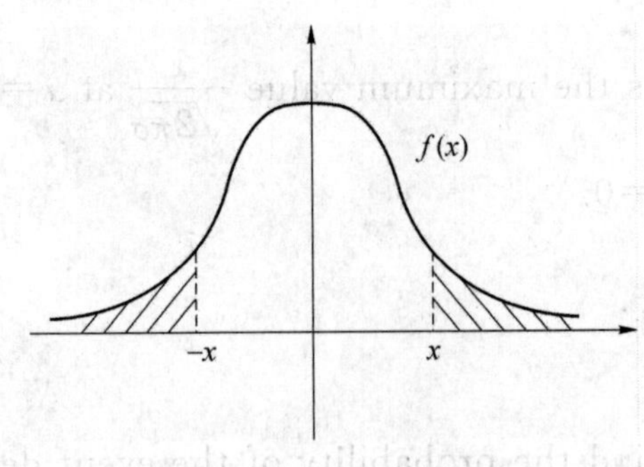

Figure 2.3.4

Example 2. 3. 9

Suppose $X \sim N\left(\frac{3}{2}, 2^2\right)$. Find the probability $P\{-1 \leqslant X \leqslant 2\}$.

Solution.

$$P\{-1 \leqslant X \leqslant 2\} = \Phi\left(\frac{2-\frac{3}{2}}{2}\right) - \Phi\left(\frac{-1-\frac{3}{2}}{2}\right) = \Phi\left(\frac{1}{4}\right) - \Phi\left(-\frac{5}{4}\right) = \Phi\left(\frac{1}{4}\right) + \Phi\left(\frac{5}{4}\right) - 1 = 0.5987 + 0.8944 - 1 = 0.4931.$$

Example 2. 3. 10

Suppose $X \sim N(2, 2^2)$. (1) Compute $P\{X^2 - X < 6\}$. (2) Find a proper a, such that $P\{X < a\} = P\{X > a\}$.

Solution.

(1) $P\{X^2 - X < 6\} = P\{-2 < X < 3\} = \Phi(0.5) - \Phi(-2) = \Phi(0.5) + \Phi(2) - 1 = 0.6687$.

(2) Suppose that $P\{X < a\} = P\{X > a\} \Leftrightarrow P\{X < a\} = 1 - P\{X < a\}$.

Then $P\{X < a\} = \frac{1}{2} \Rightarrow \Phi\left(\frac{a-2}{2}\right) = \frac{1}{2} \Rightarrow \frac{a-2}{2} = 0 \Rightarrow a = 2$.

Example 2. 3. 11 Error Principle

Suppose $X \sim N(\mu, \sigma^2)$. Consider the following probabilities of error bounds. (1) $P\{|X-\mu| \leqslant \sigma\}$. (2) $P\{|X-\mu| \leqslant 2\sigma\}$. (3) $P\{|X-\mu| \leqslant 3\sigma\}$.

Solution.

(1) $P\{|X-\mu| \leqslant \sigma\} = P\{\mu - \sigma \leqslant X \leqslant \mu + \sigma\} = \Phi(1) - \Phi(-1) = 2\Phi(1) - 1 = 0.6826$.

(2) $P\{|X-\mu| \leqslant 2\sigma\} = 2\Phi(2) - 1 = 0.9544$.

(3) $P\{|X-\mu| \leqslant 3\sigma\} = 2\Phi(3) - 1 = 0.9974$.

From the above analysis, we can see that normally distributed r. v. X almost falls in $[\mu - 3\sigma, \mu + 3\sigma]$. This is called "$3\sigma$-principle".

2.4 The Function of a Random Variable

In this section, we consider the following model:

Suppose that there exists a measurable function g such that $Y=g(X)$ for two random variables X and Y. The p. f / p. d. f of X is already known. Find the p. f. / p. d. f. of Y.

We will discuss this model in two parts.

Part 1: X is discrete.

Suppose that X is a discrete random variable with possible values $x_k, k=1,2,\cdots$ and $P\{X=x_k\}=p_k$. Denote $Y=g(X)$. We shall find the p. f. of Y.

Suppose that the range of Y is $\{y_n, n=1,2,\cdots\}$.

$$P\{Y=y_n\}=P\{g(X)=y_n\}=P\{g(X)=y_n\}$$
$$=P\{X=x_k:g(x_k)=y_n\}=\sum_{k:g(x_k)=y_n} p_k.$$

Example 2.4.1

A discrete r. v. X has the following p. f.

$X=x_k$	-1	0	1
$P\{X=x_k\}=p_k$	$\frac{1}{2}$	$\frac{1}{4}$	$\frac{1}{4}$

Compute the p. f. of (1) $Y=2X+1$, (2) $Z=X^2$.

Solution.

(1) Y takes values from $\{-1,1,3\}$.

$$P\{Y=-1\}=P\{2X+1=-1\}=P\{X=-1\}=\frac{1}{2}.$$

$$P\{Y=1\}=P\{2X+1=1\}=P\{X=0\}=\frac{1}{4}.$$

$$P\{Y=3\}=P\{2X+1=3\}=P\{X=1\}=\frac{1}{4}.$$

(2) Z takes values from $\{0,1\}$.

$$P\{Z=0\}=P\{X^2=0\}=P\{X=0\}=\frac{1}{4}.$$

$$P\{Z=1\}=P\{X^2=1\}=P\{X=-1\}+P\{X=1\}=\frac{3}{4}.$$

Part 2: X is continuous.

Suppose that X is a continuous random variable with p. d. f. $f_X(x)$. Denote $Y=g(X)$. We shall find the p. d. f. $f_Y(y)$ of Y. First we find the d. f. $F_Y(y)$ of Y. Define $A_y=\{x: g(x)\leqslant y\}$.

$$F_Y(y)=P\{Y\leqslant y\}=P\{g(X)\leqslant y\}=P\{X\in A_y\}=\int_{A_y} f_X(x)\mathrm{d}x.$$

Next, $f_Y(y)=\frac{\mathrm{d}}{\mathrm{d}y}F(y)$.

Example 2.4.2

Suppose $X\sim N(0,1)$. $Y=X^2$. Determine the p. d. f. of Y.

Solution.

First we find the d. f. $F_Y(y)$ of Y.

For $y<0$, $\{X^2\leqslant y\}=\phi$. Then $F_Y(y)=P\{Y\leqslant y\}=P\{X^2\leqslant y\}=0$.

For $y\geqslant 0$,

$$F_Y(y)=P\{Y\leqslant y\}=P\{X^2\leqslant y\}=P\{-\sqrt{y}\leqslant X\leqslant\sqrt{y}\}=\frac{1}{\sqrt{2\pi}}\int_{-\sqrt{y}}^{\sqrt{y}}\mathrm{e}^{-\frac{x^2}{2}}\mathrm{d}x.$$

Next, for y>0,

$$\begin{aligned} f_Y(y)&=\frac{\mathrm{d}}{\mathrm{d}y}F(y)=\frac{\mathrm{d}}{\mathrm{d}y}\left(\frac{1}{\sqrt{2\pi}}\int_{-\sqrt{y}}^{\sqrt{y}}\mathrm{e}^{-\frac{x^2}{2}}\mathrm{d}x\right)\\ &=\frac{1}{\sqrt{2\pi}}\mathrm{e}^{-\frac{y}{2}}\times\frac{1}{2\sqrt{y}}+\frac{1}{\sqrt{2\pi}}\mathrm{e}^{-\frac{y}{2}}\times\frac{1}{2\sqrt{y}}=\frac{1}{\sqrt{2\pi y}}\mathrm{e}^{-\frac{y}{2}}.\end{aligned}$$

In summary,

$$f_Y(y)=\begin{cases}\frac{1}{\sqrt{2\pi y}}\mathrm{e}^{-\frac{y}{2}}, & y>0;\\ 0, & y\leqslant 0.\end{cases}$$

With this p. d. f. $f_Y(y)$, we say that Y is distributed according to the chi-squared (χ^2) distribution, which is one of the most widely used probability distributions in statistics, e. g., in construction of confidence interval and or in hypothesis testing, as well as normal distribution.

Theorem 2. 4. 1

Suppose that X is a continuous random variable with p. d. f. $f_X(x)$ and d. f. $F_X(x)$. The range of X is the interval $[a,b]$. Denote $Y=g(X)$. $g(\cdot)$ is a differential function and $g'(x)>0$ for any $x\in[a,b]$ (or $g'(x)<0$ for any $x\in[a,b]$). Then the p. d. f. $f_Y(y)$ of Y is

$$f_Y(y)=\begin{cases} f_X(g^{-1}(y))\times\left|\dfrac{d}{dy}g^{-1}(y)\right|, & \alpha\leqslant y\leqslant\beta; \\ 0, & \text{otherwise}, \end{cases}$$

where $\alpha=\min\limits_{x\in[a,b]} g(x)$, $\beta=\max\limits_{x\in[a,b]} g(x)$.

Proof.

We fist find the d. f. $F_Y(y)$

Obviously, for $y<\alpha$, $\{Y\leqslant y\}=\phi$ which implies $F_Y(y)=0$.

For $y>\beta$, $\{Y\leqslant y\}=S$ which implies $F_Y(y)=1$.

For $\alpha\leqslant y\leqslant\beta$, if $g'(x)>0$

$F_Y(y)=P\{Y\leqslant y\}=P\{g(X)\leqslant y\}=P\{X\leqslant g^{-1}(y)\}=F_X(g^{-1}(y))$.

Otherwise,

$F_Y(y)=P\{Y\leqslant y\}=P\{g(X)\leqslant y\}=P\{X\geqslant g^{-1}(y)\}=1-F_X(g^{-1}(y))$.

Then we calculate the p. d. f.

If $g'(x)>0$, $\dfrac{d}{dy}g^{-1}(y)>0$. Then $\dfrac{d}{dy}g^{-1}(y)=\left|\dfrac{d}{dy}g^{-1}(y)\right|$.

$$f_Y(y)=\begin{cases} \dfrac{d}{dy}F_X(g^{-1}(y)), & \alpha\leqslant y\leqslant\beta; \\ 0, & \text{otherwise}. \end{cases}$$

$$=\begin{cases} f_X(g^{-1}(y))\left|\dfrac{d}{dy}g^{-1}(y)\right|, & \alpha\leqslant y\leqslant\beta; \\ 0, & \text{otherwise}. \end{cases}$$

If $g'(x)<0$, $\dfrac{d}{dy}g^{-1}(y)<0$. Then $-\dfrac{d}{dy}g^{-1}(y)=\left|\dfrac{d}{dy}g^{-1}(y)\right|$.

$$f_Y(y)=\begin{cases}\dfrac{d}{dy}[1-F_X(g^{-1}(y))], & \alpha\leqslant y\leqslant\beta;\\ 0, & \text{otherwise.}\end{cases}$$

$$=\begin{cases}-f_X(g^{-1}(y))\dfrac{d}{dy}g^{-1}(y), & \alpha\leqslant y\leqslant\beta;\\ 0, & \text{otherwise.}\end{cases}$$

$$=\begin{cases}f_X(g^{-1}(y))\left|\dfrac{d}{dy}g^{-1}(y)\right|, & \alpha\leqslant y\leqslant\beta;\\ 0, & \text{otherwise.}\end{cases}$$

Example 2.4.3

Suppose $X\sim N(\mu,\sigma^2)$. $Y=aX+b$. Determine the p. d. f. of Y.

Solution.

Notice that $Y=aX+b$ is a linear function. Define $y=g(x)=ax+b$. Its range is R and $g^{-1}(y)=\dfrac{y-b}{a}$.

Then

$$f_Y(y)=\frac{1}{\sqrt{2\pi}\sigma}e^{-\frac{(\frac{y-b}{a}-\mu)^2}{2\sigma^2}}\times\frac{1}{|a|}=\frac{1}{\sqrt{2\pi}|a|\sigma}e^{-\frac{(y-(a\mu+b))^2}{2(a\sigma)^2}},y\in\mathbf{R}.$$

From the above p. d. f., it is easy to see that $Y\sim N(a\mu+b,(a\sigma)^2)$.

Specially, if $a=\dfrac{1}{\sigma}$, $b=-\dfrac{\mu}{\sigma}$, i. e., $Y=g(X)=\dfrac{X-\mu}{\sigma}\sim N(0,1)$, which is consistent with Theorem 2.3.5.

Example 2.4.4

Suppose that X is a continuous random variable with d. f. $F_X(x)$ that is strictly increasing in x. Denote $Y=F_X(X)$. Show that Y is uniformly distributed in $[0,1]$.

Proof.

Compute the d. f. $F_Y(y)$ of Y.

From the definition the range of Y is $[0,1]$.

For $y<0$, $\{Y\leqslant y\}=\phi$, $F_Y(y)=0$.

For $y>1$, $\{Y\leqslant y\}=S$, $F_Y(y)=1$.

For $0\leqslant y\leqslant 1$, $P\{Y\leqslant y\}=P\{X\leqslant F_X^{-1}(y)\}=F_X(F_X^{-1}(y))=y$.

In summary,

$$F_Y(y)=\begin{cases}1, & y>1;\\ y, & 0\leqslant y\leqslant 1;\\ 0, & y<0.\end{cases}$$

Therefore, $Y\sim U(0,1)$.

Example 2.4.5

Suppose that $X\sim U(0,1)$. Use X to construct a random variable Y such that $Y\sim \text{Exp}(\lambda)$.

Solution.

The d. f. of a random variable exponentially distributed with parameter λ is

$$F_Y(y)=\begin{cases}1-e^{-\lambda y}, & y\geqslant 0;\\ 0, & \text{otherwise}.\end{cases}$$

Define $x=F_Y(y)$ for $y\geqslant 0$. Then we have $y=F_Y^{-1}(x)=-\frac{1}{\lambda}\log(1-x)$.

Define $Y=F_Y^{-1}(X)=-\frac{1}{\lambda}\log(1-X)$. It is a strictly increasing function. Then we claim that $Y\sim \text{Exp}(\lambda)$

For $y<0$, $\{Y\leqslant y\}=\phi$, $F_Y(y)=0$.

For $y\geqslant 0$, $P\{Y\leqslant y\}=P\left\{-\frac{1}{\lambda}\log(1-X)\leqslant y\right\}=P\{X\leqslant 1-e^{-\lambda y}\}=1-e^{-\lambda y}$.

Exercises

2.1 Suppose that the random variables X can take only the values $-1,0,1,2$; and that the probability function of X is specified in the following table:

$X=x_k$	-1	0	1	2
$P\{X=x_k\}=p_k$	0.1	0.2	0.4	0.3

Then find $P\{X\geqslant 1.5\}$.

2.2 Suppose that $F(x)$ and $G(x)$ are d. f. 's with $F(x)=aG(3x)+0.7G\left(\frac{x}{2}+1\right)$. Find a.

2.3 Suppose that $P\{X=k\}=\frac{1}{2^k}$, $k=1,2,\cdots$ Compute: (1) $P\{X \text{ is even}\}$;

(2) $P\{X\geqslant 5\}$.

2.4 Compute $P\{X \text{ is even}\}$: (1) if $X\sim B(n,p)$; (2) if $X\sim P(\lambda)$.

2.5 Suppose that $X\sim B(2,p)$, $Y\sim B(3,p)$. $P\{X=0\}=\frac{4}{9}$. Find $P\{Y\geqslant 1\}$.

2.6 Suppose that $X\sim B(n,p)$. Find k, such that $P\{X=k\}=\max\limits_{i=1,2,\cdots,n} P\{X=i\}$.

2.7 Suppose that $X\sim P(\lambda)$. Find k, such that $P\{X=k\}=\max\limits_{i=1,2,\cdots} P\{X=i\}$.

2.8 Suppose that X has a Poisson distribution with parameter λ. We have known that $P\{X=1\}=P\{X=2\}$. Then find $P\{X\geqslant 1\}$.

2.9 One number is selected from the set $\{1,2,3,4\}$ at random. The selected number is denoted by X. Then another number is selected from a new set $\{1,2,\cdots,X\}$ at random, denoted by Y. Compute $P\{Y=2\}$.

2.10 Suppose that the d. f. of random variable X is as follows:

$$F(x)=\begin{cases} 0, & x<-1; \\ 1/8, & x=-1; \\ ax+b, & -1<x<1; \\ 1, & x\geqslant 1. \end{cases}$$

$P\{X=1\}=1/4$. (1) Find a and b. (2) Determine $P\{X<1\}$ and $P\{-1\leqslant X\leqslant 1\}$.

2.11 Suppose that the d. f. of random variable X is as follows:

$$F(x)=\begin{cases} \frac{1}{2}e^{x}, & x<0; \\ 1-\frac{1}{2}e^{-x}, & x\geqslant 0. \end{cases}$$

Then we shall determine the p. d. f. of X. Compute $P\{0<X<2\}$.

2.12 Suppose that the p. d. f. of r. v. X is specified as follows:

$$f(x)=\begin{cases} 2e^{-2x}, & x>0; \\ 0, & x\leqslant 0. \end{cases}$$

Determine $F(5)$.

2.13 Suppose that the d. f. of continuous random variable X is as follows:

$$F(x)=\begin{cases} 0, & x<0; \\ ax^2, & 0\leqslant x<2; \\ 1, & x\geqslant 2. \end{cases}$$

Find a.

2.14 Suppose that the p. d. f. of the random variable X is defined as follows:

$$f(x)=\begin{cases} \frac{1}{2}\sin x, & 0<x<\pi; \\ 0, & \text{otherwise}. \end{cases}$$

Find the d. f.

2.15 Let $N(t)$ be the number of cars passing through some intersection in time $(0,t]$. Suppose that $N(t)\sim P(\lambda t)$, where λ is a positive constant. The inter-arrival times of those cars are denoted by $T_1, T_2, \cdots$. Compute the p. d. f. of T_1.

2.16 Suppose that some bolts are produced by one machine. The length (cm) of the bolts are distributed as $N(10, 0.06^2)$. A bolt is regarded as qualified if the range of the length is between $(10-0.12)$ cm and $(10+0.12)$ cm. Now a bolt is sampled at random. Determine the probability that it is not qualified.

2.17 Suppose that X has an exponential distribution with parameter 1. Determine the p. d. f. of $Y=X^2$.

2.18 Suppose that X has a normal distribution with parameters 2 and 4. Find the p. d. f. of $Y=1-X$.

2.19 Suppose that the p. d. f. of random variable X is as follows:

$$f(x)=\frac{1}{2}e^{-|x|}, \quad x\in \mathbf{R}.$$

Find the p. d. f of r. v. Y: (1) $Y=|X|$; (2) $Y=\log|X|$.

2.20 Suppose that $X\sim N(0,1)$. Find the p. d. f of r. v. $Y=e^X$.

Chapter 3 Multi-Dimensional Random Variable and Distributions

So far, we have only considered probability distributions for single random variable. However, we are often faced by probability statements concerning two or more random variables. For example, inspect the height and weight of students in the first grade in some elementary school. We let all the students in the first grade in the elementary school be the sample space $S=\{s\}$, let $H=H(s)$ be the height and $W=W(s)$ be the weight of any student. In order to find the relationship between height and weight, we often study H and W together, that is, (H,W) taken as a whole, which is a two-dimensional random variable.

In this chapter, we mainly focus on the two-dimensional random variable and its probability distribution.

3.1 Multi-Dimensional Random Variable and its Distribution

In general, the joint distribution of more than two random variables is called a joint multivariate distribution, and of two random variables is called bivariate distribution.

3.1.1 Multi-Dimensional Random Variable and its Multivariate Distribution

Definition 3.1.1

Suppose that $X_1(s), X_2(s), \cdots, X_n(s)$ are n random variables defined on the probability space $(S, \mathscr{F}, P)$, then the n-dimensional vector $(X_1(s), X_2(s), \cdots, X_n(s))$ is called as the n-dimensional random vector or the n-dimensional random variable, simply denoted by $(X_1, X_2, \cdots, X_n)$, where X_i is called as the i-th element, $i=1,2,\cdots,n$.

By Definition 3.1.1, for any given sample $(X_1, X_2, \cdots, X_n) \in \mathbf{R}^n$. like one-dimensional random variable, we use the distribution function to study the statistical law of $(X_1, X_2, \cdots, X_n)$.

Definition 3.1.2

Suppose that $(X_1(s), X_2(s), \cdots, X_n(s))$ is a n-dimensional random variable. We define the joint cumulative distribution function (c. d. f.):

$$F(x_1, x_2, \cdots, x_n) = P\{X_1 \leqslant x_1, X_2 \leqslant x_2, \cdots, X_n \leqslant x_n\}. \qquad (3.1.1)$$

Sometimes F is called as distribution function (d. f.) of $(X_1, X_2, \cdots, X_n)$.

We note that $\{X_1 \leqslant x_1, X_2 \leqslant x_2, \cdots, X_n \leqslant x_n\}$ is a product event, that is

$$\{X_1 \leqslant x_1, X_2 \leqslant x_2, \cdots, X_n \leqslant x_n\} = \prod_{i=1}^{n} \{X_i \leqslant x_i\}.$$

From now, we mainly focus on the case $n=2$, that is, two-dimensional random variable (X, Y). Most of properties and results on (X, Y) can be similarly generalized to the multi-dimensional random variable.

3.1.2 Joint Distribution Function of (X, Y)

The joint d. f., of two random variables X and Y is defined as the function F such that for all values of x and y,

$$F(x, y) = P\{X \leqslant x \text{ and } Y \leqslant y\}, \qquad (3.1.2)$$

The pvobability $F(x, y)$ in (3.1.2) is sketched in Figure 3.1.1.

where $-\infty < x < \infty$ and $-\infty < y < \infty$.

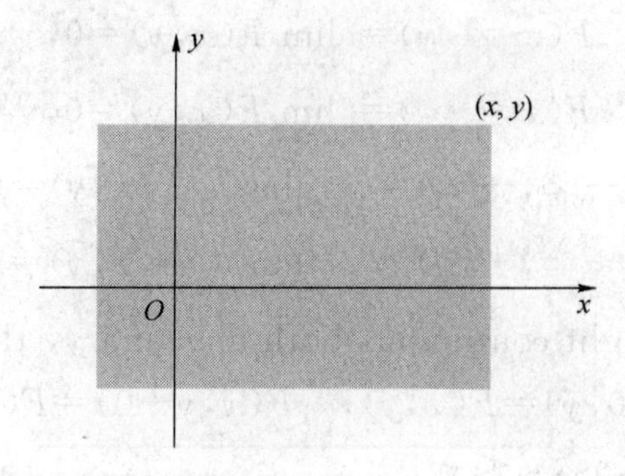

Figure 3. 1. 1

If the joint d. f. of two arbitrary random variables X and Y is F, then the probability that the pair (X, Y) will lie in a specified rectangle in the xy-plane can be found from F as follows: For any given numbers $x_1 < x_2$, $y_1 < y_2$,

$$\begin{aligned} &P\{x_1 < X \leqslant x_2 \text{ and } y_1 < Y \leqslant y_2\} \\ &= P\{x_1 < X \leqslant x_2 \text{ and } Y \leqslant y_2\} - P\{x_1 < X \leqslant x_2 \text{ and } Y \leqslant y_1\} \\ &= P\{X \leqslant x_2 \text{ and } Y \leqslant y_2\} - P\{X \leqslant x_1 \text{ and } Y \leqslant y_2\} \\ &\quad - P\{X \leqslant x_2 \text{ and } Y \leqslant y_1\} + P\{X \leqslant x_1 \text{ and } Y \leqslant y_1\} \\ &= F(x_2, y_2) - F(x_1, y_2) - F(x_2, y_1) + F(x_1, y_1). \end{aligned} \quad (3.1.3)$$

The probability in formula (3. 1. 3) is sketched in Figure 3. 1. 2.

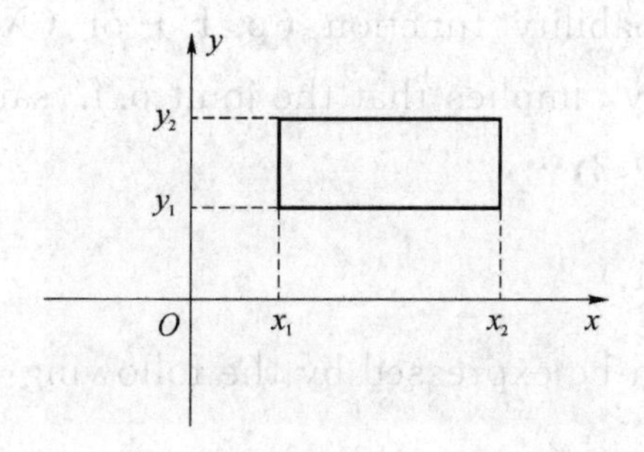

Figure 3. 1. 2

The joint d. f. $F(x, y)$ has the following properties:

(1) Fix x, $F(x, y)$ is nondecreasing as y increases. Similarly, fix y, $F(x, y)$ is nondecreasing as x increases.

(2) For any x and y, $0 \leqslant F(x, y) \leqslant 1$ and

$$F(-\infty, y) = \lim_{x \to -\infty} F(x, y) = 0,$$

$$F(x, -\infty) = \lim_{y \to -\infty} F(x, y) = 0,$$

$$F(-\infty, -\infty) = \lim_{x \to -\infty, y \to -\infty} F(x, y) = 0,$$

$$F(+\infty, +\infty) = \lim_{x \to +\infty, y \to +\infty} F(x, y) = 1.$$

(3) $F(x, y)$ is right continuous both in x and y, that is,

$$F(x+0, y) = F(x, y), \quad F(x, y+0) = F(x, y).$$

(4) For any $x_1 \leqslant x_2, y_1 \leqslant y_2$,

$$F(x_2, y_2) - F(x_1, y_2) - F(x_2, y_1) + F(x_1, y_1) \geqslant 0.$$

The proofs of properties (1) (2) and (3) are similar with the one-dimensional d. f., and the proof of (4) follows from formula (3.1.3).

3.1.3 Discrete Bivariate Distribution

Definition 3.1.3

It is said that the random variable (X, Y) has a discrete distribution or that (X, Y) is a discrete bivariate random variable if (X, Y) can take values (x_i, y_j), $i, j = 1, 2, \cdots$.

If (X, Y) takes (x_i, y_j), $i = 1, 2, \cdots, j = 1, 2, \cdots$, then

$$P\{X = x_i, Y = y_j\} = p_{ij}, \quad i, j = 1, 2, \cdots$$

is called as joint probability function (p. f.) of (X, Y). This, and the definition of probability, implies that the joint p. f. satisfies that

(1) $p_{ij} \geqslant 0, i, j = 1, 2, \cdots,$

(2) $\sum_{i=1}^{\infty} \sum_{j=1}^{\infty} p_{ij} = 1.$

The joint p. f. can be expressed by the following Table 3.1.1.

Table 3. 1. 1

X \ Y	y_1	y_2	…	y_n	…
x_1	p_{11}	p_{21}	…	p_{1n}	…
x_2	p_{21}	p_{22}	…	p_{2n}	…
⋮	⋮	⋮		⋮	
x_m	p_{m1}	p_{m2}	…	p_{mn}	…
⋮	⋮	⋮		⋮	

Example 3. 1. 1

Suppose that the random variable X can take only the values 1, 2 and 3; that the random variable Y can take only the values 1, 2, 3 and 4; and that the joint p. f. of X and Y is as specified in the following Table 3. 1. 2.

Table 3. 1. 2

X \ Y	1	2	3	4
1	0. 1	0	0. 1	0
2	0. 3	0	0. 1	0. 2
3	0	0. 2	0	0

Determine the values of $P\{X\geqslant 2 \text{ and } Y\geqslant 2\}$ and $P\{X=1\}$.

Solution.

By summing p_{ij} over all values of $x\geqslant 2$ and $y\geqslant 2$, we obtain the value

$$P\{X\geqslant 2 \text{ and } Y\geqslant 2\}=\sum_{i\geqslant 2, j\geqslant 2} p_{ij}=p_{23}+p_{24}+p_{32}=0.5.$$

By summing the probabilities in the first row of the table, we obtain the value

$$P\{X=1\}=\sum_{j=1}^{4} p_{1j}=0.2.$$

3.1.4 Continuous Bivariate Distribution

Definition 3.1.4

It is said that a bivariate random variable (X,Y) has a continuous joint distribution or that (X,Y) is a continuous random variable if there exists a nonnegative function $f(x,y)$ defined over the entire xy-plane such that for any x,y, the joint d. f.

$$F(x,y) = \int_{-\infty}^{x}\int_{-\infty}^{y} f(u,v)\,du\,dv.$$

The function f is called the joint probability density function (p. d. f.) of X and Y.

The joint p. d. f. $f(x,y)$ must satisfy the following conditions:

(1) $f(x,y)\geqslant 0$ for $-\infty<x<\infty, -\infty<y<\infty$.

(2) $\int_{-\infty}^{\infty}\int_{-\infty}^{\infty} f(x,y)\,dx\,dy = 1$.

(3) If f is continuous at (x,y), then

$$\frac{\partial^2 F(x,y)}{\partial x\partial y}=\frac{\partial^2 F(x,y)}{\partial y\partial x}=f(x,y).$$

(4) Suppose D is a specified region in the xy-plane, then

$$P\{(X,Y)\in D\} = \iint_{(x,y)\in D} f(x,y)\,dx\,dy.$$

That is, the probability that the pair (X,Y) will belong to each specified region of the xy-plane can be found by integrating the joint p. d. f. $f(x,y)$ over that region.

If (X,Y) has a continuous joint distribution, then (i) every individual point and every infinite sequence of points in the xy-plane has probability 0. (ii) every one-dimensional curve in the xy-plane has probability 0. Thus, the probability that (X,Y) lies on each specified straight line in the plane is 0, and the probability that (X,Y) lies on each specified circle in the plane is 0.

Example 3.1.2

Suppose (X,Y) is a continuous bivariate random variable, its p. d. f.

$$f(x,y)=\begin{cases} Ax, & 0<x<1, 0<y<x; \\ 0, & \text{otherwise.} \end{cases}$$

Find (1) the constant A; (2) $P\left\{X>\frac{3}{4}\right\}$; (3) $P\left\{X<\frac{1}{4},Y<\frac{1}{2}\right\}$.

Solution.

(1) By the property of the p. d. f. ,

$$1=\int_{-\infty}^{\infty}\int_{-\infty}^{\infty} f(x,y)\mathrm{d}x\mathrm{d}y=\int_{0}^{1}\int_{0}^{x} Ax\,\mathrm{d}y\mathrm{d}x=\frac{A}{3},$$

then $A=3$.

(2) Let $D=\left\{(x,y):x>\frac{3}{4}\right\}$, and $\left\{X>\frac{3}{4}\right\}=\{(X,Y)\in D\}$, then we have

$$P\left\{X>\frac{3}{4}\right\}=P\{(X,Y)\in D\}=\iint\limits_{(x,y)\in D}(x,y)\mathrm{d}x\mathrm{d}y=\int_{\frac{3}{4}}^{1}\int_{0}^{x}3x\mathrm{d}y\mathrm{d}x=\frac{37}{64}.$$

(3) Let $E=\{(x,y):0<x<\frac{1}{4},0<y<\frac{1}{2}\}$, similarly with (2) above,

$$P\left\{X<\frac{1}{4},Y<\frac{1}{2}\right\}=P\{(X,Y)\in E\}=\int_{0}^{\frac{1}{4}}\int_{0}^{x}3x\mathrm{d}y\mathrm{d}x=\frac{1}{64}.$$

Example 3. 1. 3

Suppose that the joint p. d. f. of X and Y is specified as follows

$$f(x,y)=\begin{cases} cx^2y, & x^2\leqslant y\leqslant 1;\\ 0, & \text{otherwise},\end{cases}$$

for some constant c. Determine (1) c; (2) $P\{X\geqslant Y\}$.

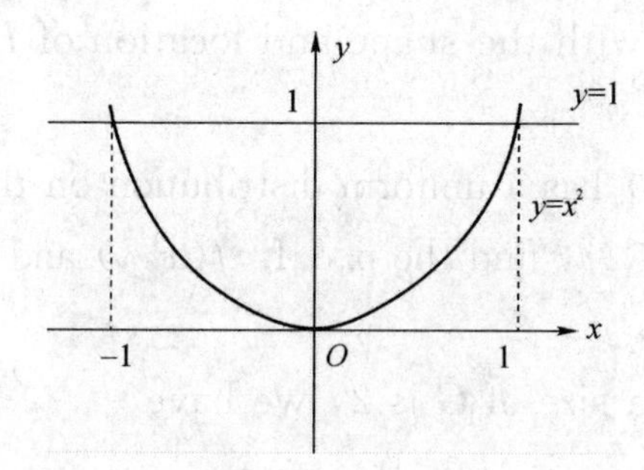

Figure 3. 1. 3

Solution.

(1) We first plot the area $x^2\leqslant y\leqslant 1$ in Figure 3. 1. 3. By the property of the p. d. f. , it follows that

$$1=\int_{-\infty}^{\infty}\int_{-\infty}^{\infty} f(x,y)\mathrm{d}x\mathrm{d}y=\iint_S f(x,y)\mathrm{d}x\mathrm{d}y=\int_{-1}^{1}\int_{x^2}^{1} cx^2 y\mathrm{d}y\mathrm{d}x=\frac{4}{21}c.$$

Since the value of this integral must be 1, the value of c must be 21/4.

(2) Since $\{X\geqslant Y\}=\{(X,Y)\in\{(x,y):x\geqslant y\}\}$,

$$P(X\geqslant Y)=\iint_{x\geqslant y} f(x,y)\mathrm{d}x\mathrm{d}y=\int_0^1\int_{x^2}^{x}\frac{21}{4}x^2 y\mathrm{d}y\mathrm{d}x=\frac{3}{20}.$$

3.1.5 Two special distributions

Definition 3.1.5 (Uniform Distribution)

Suppose that G is a bounded region with the area size S in the xy-plane. If (X,Y) has the following p. d. f.

$$f(x,y)=\begin{cases}\frac{1}{S}, & (x,y)\in G;\\ 0, & \text{otherwise},\end{cases}$$

then (X,Y) is said to have a uniform distribution.

If (X,Y) has a uniform distribution on region G, D is a subregion located in G and has the area size $S(D)$, then

$$P\{(X,Y)\in D\}=\iint_{(x,y)\in D} f(x,y)\mathrm{d}x\mathrm{d}y=\iint_{(x,y)\in D}\frac{1}{S}\mathrm{d}x\mathrm{d}y=\frac{S(D)}{S},$$

which is nothing to do with the shape and location of D.

Example 3.1.4

Suppose that (X,Y) has a uniform distribution on the rectangle region $G=\{(x,y):0\leqslant x\leqslant 1,0\leqslant y\leqslant 2\}$, find the p. d. f. $f(x,y)$ and the d. f. $F(x,y)$.

Solution.

(Ⅰ) Since the area size of G is 2, we have

$$f(x,y)=\begin{cases}\frac{1}{2}, & 0\leqslant x\leqslant 1,0\leqslant y\leqslant 2;\\ 0, & \text{otherwise}.\end{cases}$$

(Ⅱ) Notice that for any x,y,

$$F(x,y)=\int_{-\infty}^{x}\int_{-\infty}^{y} f(u,v)\mathrm{d}u\mathrm{d}v.$$

We have

(1) if $x<0$ or $y<0$, then $F(x,y)=0$;

(2) if $0\leqslant x<1, 0\leqslant y<2$, then $F(x,y)=\int_0^x\int_0^y \frac{1}{2}dudv=\frac{1}{2}xy$;

(3) if $x\geqslant 1, 0\leqslant y<2$, then $F(x,y)=\int_0^1\int_0^y \frac{1}{2}dudv=\frac{1}{2}y$;

(4) if $0\leqslant x<1, y\geqslant 2$, then $F(x,y)=\int_0^x\int_0^2 \frac{1}{2}dudv=x$;

(5) if $x\geqslant 1, y\geqslant 2$, then $F(x,y)=\int_0^1\int_0^2 \frac{1}{2}dudv=1$.

Hence, the d. f.

$$F(x,y)=\begin{cases} 0, & x<0 \text{ or } y<0; \\ \frac{1}{2}xy, & 0\leqslant x<1, 0\leqslant y<2; \\ \frac{1}{2}y, & x\geqslant 1, 0\leqslant y<2; \\ x, & 0\leqslant x<1, y\geqslant 2; \\ 1, & x\geqslant 1, y\geqslant 2. \end{cases}$$

Definition 3. 1. 6. (Bivariate Normal Distribution)

If (X,Y) has p. d. f. as

$$f(x,y)=\frac{1}{2\pi\sigma_1\sigma_2\sqrt{1-\rho^2}}\exp\left\{-\frac{1}{2(1-\rho^2)}\left[\left(\frac{x-\mu_1}{\sigma_1}\right)^2-2\rho\left(\frac{x-\mu_1}{\sigma_1}\right)\left(\frac{y-\mu_2}{\sigma_2}\right)+\left(\frac{y-\mu_2}{\sigma_2}\right)^2\right]\right\}, \tag{3.1.4}$$

where $-\infty<\mu_1<+\infty, -\infty<\mu_2<+\infty, \sigma_1>0, \sigma_2>0$ and $|\rho|\leqslant 1$, then (X,Y) is said to have a bivariate normal distribution, denoted by $(X,Y)\sim N(\mu_1,\mu_2,\sigma_1^2,\sigma_2^2,\rho)$.

The p. d. f. $f(x,y)$ in formula (3. 1. 4) has the following function graphs:

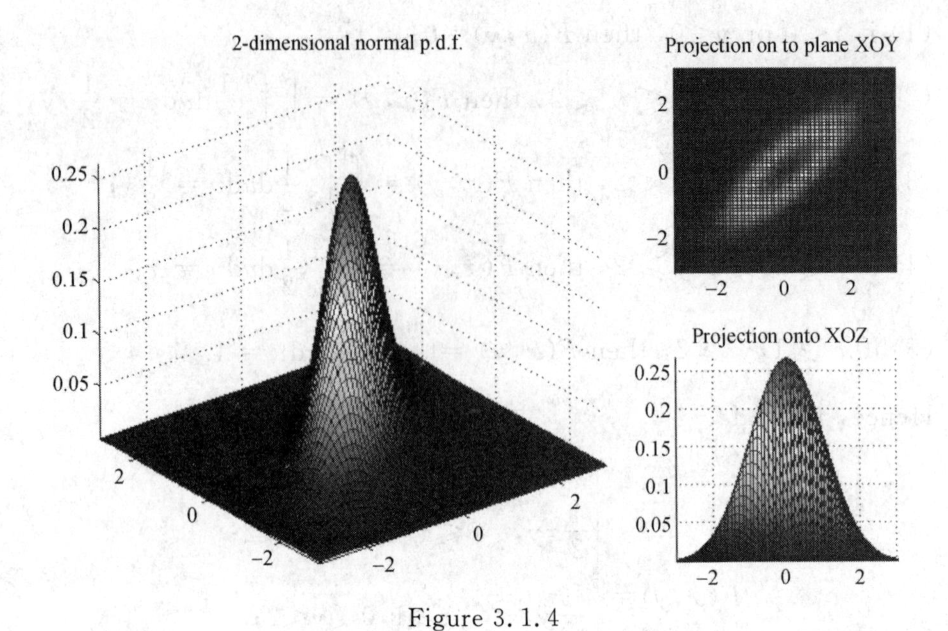

Figure 3. 1. 4

3.2 Marginal Distribution

The distribution of one random variable X computed from a joint distribution of (X,Y) is called the marginal distribution of X, so it does for Y. Each random variable will have a marginal d. f. as well as a marginal p. d. f. for continuous random variable or p. f. for discrete random variable.

3.2.1 Marginal Distribution Function

The d. f. s of X and Y computed from the joint d. f. $F(x,y)$ are called as the marginal d. f. s of X and Y respectively, and denoted by $F_X(x)$ and $F_Y(y)$. Given a joint d. f. $F(x,y)$, the marginal d. f.

$$\begin{aligned} F_X(x) &= P\{X\leqslant x\} = P\{X\leqslant x, Y<+\infty\} \\ &= \lim_{y\to+\infty} P\{X\leqslant x, Y\leqslant y\} \\ &= \lim_{y\to+\infty} F(x,y) \\ &= F(x,+\infty), \end{aligned} \tag{3.2.1}$$

$$
\begin{aligned}
F_Y(y) &= P\{Y \leqslant y\} = P\{X \leqslant +\infty, Y \leqslant y\} \\
&= \lim_{x \to +\infty} P\{X \leqslant x, Y \leqslant y\} \\
&= \lim_{x \to +\infty} F(x, y) \\
&= F(+\infty, y).
\end{aligned}
\tag{3.2.2}
$$

Example 3.2.1

Suppose that (X, Y) has joint d. f.

$$F(x, y) = A\left(B + \arctan\frac{x}{2}\right)\left(C + \arctan\frac{y}{3}\right)$$

for constants A, B and C. Find (1) A, B, C; (2) $F_X(x)$ and $F_Y(y)$; (3) $P\{0 < X \leqslant 2, 0 < Y \leqslant 3\}$.

Solution.

(1) Since

$$F(+\infty, +\infty) = A\left(B + \frac{\pi}{2}\right)\left(C + \frac{\pi}{2}\right) = 1,$$

$$F(-\infty, y) = A\left(B - \frac{\pi}{2}\right)\left(C + \arctan\frac{y}{3}\right) = 0,$$

$$F(x, -\infty) = A\left(B + \arctan\frac{x}{2}\right)\left(C - \frac{\pi}{2}\right) = 0,$$

we have

$$A = \frac{1}{\pi^2}, \quad B = \frac{\pi}{2}, \quad C = \frac{\pi}{2}.$$

(2) By formula (3.2.1) and formula (3.2.2), we have

$$F_X(x) = F(x, +\infty) = \frac{1}{\pi^2}\left(\frac{\pi}{2} + \arctan\frac{x}{2}\right) \cdot \pi = \frac{1}{2} + \frac{1}{\pi}\arctan\frac{x}{2},$$

$$F_Y(y) = F(+\infty, y) = \frac{1}{\pi^2} \cdot \pi\left(\frac{\pi}{2} + \arctan\frac{y}{3}\right) = \frac{1}{2} + \frac{1}{\pi}\arctan\frac{y}{3}.$$

(3) It follows from formula (3.1.3) that

$$
\begin{aligned}
P\{0 < X \leqslant 2, 0 < Y \leqslant 3\} &= F(2, 3) - F(0, 3) - F(2, 0) + F(0, 0) \\
&= \frac{9}{16} - \frac{3}{8} - \frac{3}{8} + \frac{1}{4} = \frac{1}{16}.
\end{aligned}
$$

3.2.2 Discrete Marginal Distribution

Suppose that (X,Y) is a bivariate random variable with p. f.

$$P\{X=x_i,Y=y_j\}=p_{ij}, \quad i,j=1,2,\cdots. \tag{3.2.3}$$

Because that

$$\begin{aligned}\{X=x_i\} &= \{X=x_i\} \cap S \\ &= \{X=x_i\} \cap (\bigcup_{j=1}^{\infty} \{Y=y_j\}) \\ &= \bigcup_{j=1}^{\infty} (\{X=x_i\} \cap \{Y=y_j\}) \\ &= \bigcup_{j=1}^{\infty} \{X=x_i, Y=y_j\},\end{aligned}$$

where $\{X=x_i, Y=y_j\}$ is disjoint event sequence for all i, j. So, by the definition of probability, for $i=1,2,\cdots$,

$$P\{X=x_i\} = \sum_{j=1}^{\infty} P\{X=x_i,Y=y_j\} = \sum_{j=1}^{\infty} p_{ij} \equiv p_{i+}, \tag{3.2.4}$$

which satisfies

$$p_{i+} \geqslant 0, j=1,2,\cdots, \text{ and } \sum_{i=1}^{\infty} p_{i+} = \sum_{i=1}^{\infty}\sum_{j=1}^{\infty} p_{ij} = 1. \tag{3.2.5}$$

By formula (3.2.4) and formula (3.2.5), we know that $\{p_{i+}, i=1,2,\cdots\}$ is the p. f. of X, which is called as the marginal p. f. of X with respect to the bivariate (X,Y), or simply the marginal p. f. of X.

Similarly, the marginal p. f. of Y with respect to the bivariate (X,Y) is

$$P\{Y=y_j\} = \sum_{i=1}^{\infty} P\{X=x_i,Y=y_j\} = \sum_{i=1}^{\infty} p_{ij} \equiv p_{+j} \tag{3.2.6}$$

for $j=1,2,\cdots$.

In general, the relationship between the marginal p. f. s p_{i+}, p_{+j} and the joint p. f. p_{ij} can be shown by the following Table 3.2.1.

Table 3. 2. 1

Y \ X	y_1	y_2	…	y_n	…	p_{i+}
x_1	p_{11}	p_{21}	…	p_{1n}	…	p_{1+}
x_2	p_{21}	p_{22}	…	p_{2n}	…	p_{2+}
⋮	⋮	⋮		⋮		⋮
x_m	p_{m1}	p_{m2}	…	p_{mn}	…	p_{m+}
⋮	⋮	⋮		⋮		⋮
p_{+j}	p_{+1}	p_{+2}	…	p_{+n}	…	1

Example 3. 2. 2

Suppose that the bivariate random variable (X, Y) has a joint p. f. in Example 3. 1. 1, find the marginal p. f. of X and Y.

Solution.

With formula (3. 2. 4) and formula (3. 2. 6), it follows from Table 3. 1. 4 that

$$p_{1+} = \sum_{j=1}^{4} p_{1j} = 0.1 + 0.1 = 0.2,$$

$$p_{2+} = \sum_{j=1}^{4} p_{2j} = 0.3 + 0.1 + 0.2 = 0.6,$$

$$p_{3+} = \sum_{j=1}^{4} p_{3j} = 0.2,$$

and

$$p_{+1} = \sum_{i=1}^{3} p_{i1} = 0.1 + 0.3 = 0.4,$$

$$p_{+2} = \sum_{i=1}^{3} p_{i2} = 0.2,$$

$$p_{+3} = \sum_{i=1}^{3} p_{i3} = 0.1 + 0.1 = 0.2,$$

$$p_{+4} = \sum_{i=1}^{3} p_{i4} = 0.2.$$

3.2.3 Continuous Marginal Distribution

Suppose that the continuous bivariate random variable (X,Y) has a joint p. d. f. $f(x,y)$, we have

$$F_X(x)=F(x,+\infty)=\int_{-\infty}^{x}\left(\int_{-\infty}^{+\infty}f(u,v)\mathrm{d}v\right)\mathrm{d}u,$$

then, the p. d. f. of X is

$$f_X(x)=\frac{\mathrm{d}F_X(x)}{\mathrm{d}x}=\int_{-\infty}^{+\infty}f(x,y)\mathrm{d}y. \tag{3.2.7}$$

Similarly, we have

$$F_Y(y)=F(+\infty,y)=\int_{-\infty}^{y}\left(\int_{-\infty}^{+\infty}f(u,v)\mathrm{d}u\right)\mathrm{d}v,$$

$$f_Y(y)=\frac{\mathrm{d}F_Y(y)}{\mathrm{d}y}=\int_{-\infty}^{+\infty}f(x,y)\mathrm{d}x. \tag{3.2.8}$$

We call the $f_X(x)$ and $f_Y(y)$ the marginal p. d. f. of X and Y with respect to (X,Y), or simply the marginal p. d. f. of X and Y, respectively.

Example 3.2.3

Suppose that the bivariate (X,Y) has a joint p. d. f. $f(x,y)$ given in Example 3.1.3. Determine the marginal p. d. f. s of X and Y.

Solution.

It is said that X cannot take any value outside the interval $[-1,1]$. Therefore, $f_X(x)=0$ for $x<-1$ or $x>1$. Furthermore, for $-1\leqslant x\leqslant 1$, it can be seen that $f(x,y)=0$ unless $x^2\leqslant y\leqslant 1$. Therefore, for $-1\leqslant x\leqslant 1$,

$$f_X(x)=\int_{-\infty}^{+\infty}f(x,y)\mathrm{d}y=\int_{x^2}^{1}\frac{21}{4}x^2y\mathrm{d}y=\frac{21}{8}x^2(1-x^4).$$

Hence,

$$f_X(x)=\begin{cases}\frac{21}{8}x^2(1-x^4), & -1\leqslant x\leqslant 1;\\ 0, & \text{otherwise.}\end{cases}$$

Next, it is said that Y cannot take any value outside the interval $[0,1]$. Therefore, $f_Y(y)=0$ for $y<0$ or $y>1$. Furthermore, for $0\leqslant y\leqslant 1$, it can be seen that $f(x,y)=0$ unless $-\sqrt{y}\leqslant x\leqslant\sqrt{y}$. Therefore, for $0\leqslant y\leqslant 1$,

$$f_Y(y) = \int_{-\infty}^{\infty} f(x,y)\,\mathrm{d}x = \int_{-\sqrt{y}}^{\sqrt{y}} \frac{21}{4}x^2 y\,\mathrm{d}x = \frac{7}{2}y^{5/2}.$$

Hence,

$$f_Y(y)=\begin{cases}\frac{7}{2}y^{5/2}, & 0\leqslant y\leqslant 1,\\ 0, & \text{otherwise.}\end{cases}$$

Example 3. 2. 4

Suppose that the bivariate $(X,Y)\sim N(\mu_1,\mu_2,\sigma_1^2,\sigma_2^2,\rho)$, whose joint p. d. f. is given in formula (3. 1. 4). Find the marginal p. d. f. s of X and Y.

Solution.

Since

$$f(x)=\frac{1}{2\pi\sigma_1\sigma_2\sqrt{1-\rho^2}}\exp\left\{-\frac{1}{2(1-\rho^2)}\left[\left(\frac{x-\mu_1}{\sigma_1}\right)^2-2\rho\left(\frac{x-\mu_1}{\sigma_1}\right)\left(\frac{y-\mu_2}{\sigma_2}\right)+\left(\frac{y-\mu_2}{\sigma_2}\right)^2\right]\right\},$$

$$f_X(x) = \int_{-\infty}^{+\infty} f(x,y)\,\mathrm{d}y,$$

and

$$\left(\frac{x-\mu_1}{\sigma_1}\right)^2-2\rho\left(\frac{x-\mu_1}{\sigma_1}\right)\left(\frac{y-\mu_2}{\sigma_2}\right)+\left(\frac{y-\mu_2}{\sigma_2}\right)^2$$
$$=\left(\frac{y-\mu_2}{\sigma_2}-\rho\frac{x-\mu_1}{\sigma_1}\right)^2+(1-\rho^2)\frac{(x-\mu_1)^2}{\sigma_1^2},$$

we have

$$f_X(x) = \frac{1}{2\pi\sigma_1\sigma_2\sqrt{1-\rho^2}}\mathrm{e}^{-\frac{(x-\mu_1)^2}{2\sigma_1^2}}\int_{-\infty}^{+\infty}\mathrm{e}^{-\frac{1}{2(1-\rho^2)}\left(\frac{y-\mu_2}{\sigma_2}-\rho\frac{x-\mu_1}{\sigma_1}\right)^2}\mathrm{d}y.$$

Let

$$t=\frac{1}{\sqrt{1-\rho^2}}\left(\frac{y-\mu_2}{\sigma_2}-\rho\frac{x-\mu_1}{\sigma_1}\right),$$

it follows that

$$f_X(x) = \frac{1}{2\pi\sigma_1}\mathrm{e}^{-\frac{(x-\mu_1)^2}{2\sigma_1^2}}\int_{-\infty}^{+\infty}\mathrm{e}^{-\frac{t^2}{2}}\mathrm{d}t = \frac{1}{\sqrt{2\pi}\sigma_1}\mathrm{e}^{-\frac{(x-\mu_1)^2}{2\sigma_1^2}}.$$

Similarly,

$$f_Y(y)=\frac{1}{\sqrt{2\pi}\sigma_2}\mathrm{e}^{-\frac{(y-\mu_2)^2}{2\sigma_2^2}}.$$

From Example 3.2.4, it is easy to say that $X \sim N(\mu_1, \sigma_1^2)$, $Y \sim N(\mu_2, \sigma_2^2)$, that is, both the marginal distributions of bivariate normal distribution are one-dimensional normal distributions, and is nothing to do with the parameter ρ. For given parameters μ_1, σ_1^2, μ_2, σ_2^2, the bivariate normal distribution depends on parameter ρ, however the marginal distributions are independent of ρ. It shows that the marginal distributions can not generally determine the joint distribution. Specially, it is known that both X and Y have normal distributions, we do not know the distribution of (X, Y).

3.3 Conditional Distribution

We generalize the concept of conditional probability to conditional distribution. When an event B occurs and affects the probability that another event A occurs, we introduce the conditional probability. Similarly, when a random variable X takes values, it may affect the distribution of another random variable Y. At this time, we introduce conditional distribution. Generally, conditional distribution follows the following form: for two random variables X and Y, we find the distribution of X given that Y takes a special value. For example, consider all male students in a high school, let X and Y denote the weight and height for any male student, then X and Y are two random variables with respective distributions. Now $Y = 170$ cm is given, we want to find the distribution of X. It is reasonable to think that the distributions of X with and without condition $Y=170$ cm are different. Since the distribution of X varies accordingly to the values taken by Y, like a function of the value of Y, the conditional distribution is a research content for us.

3.3.1 Discrete Conditional Distribution

Suppose that (X, Y) is bivariate discrete random variable with joint p. f.

$$P\{X=x_i, Y=y_j\}=p_{ij}, \quad i,j=1,2,\cdots,$$

and the marginal p. f. s

$$p_{i+} = P\{X = x_i\} = \sum_{j=1}^{\infty} p_{ij}, \quad i = 1,2,\cdots,$$

$$p_{+j} = P\{Y = y_j\} = \sum_{i=1}^{\infty} p_{ij}, \quad j = 1,2,\cdots.$$

For given j such that $p_{+j}>0$, we want to find the conditional probability of event $\{X=x_i\}$ given $\{Y=y_j\}$, denoted by $P\{X=x_i \mid Y=y_j\}$. By the definition of conditional probability, we have

$$P\{X=x_i \mid Y=y_j\}=\frac{P\{X=x_i, Y=y_j\}}{P\{Y=y_j\}}=\frac{p_{ij}}{p_{+j}},$$

which has two properties:

(1) $P\{X=x_i \mid Y=y_j\}\geqslant 0, i=1,2,\cdots,$

(2) $\sum_{i=1}^{\infty} P\{X = x_i \mid Y = y_j\} = \sum_{i=1}^{\infty} \frac{p_{ij}}{p_{+j}} = 1.$

Next we introduce the following definition of the conditional distribution.

Definition 3.3.1 (conditional distribution)

Suppose that (X,Y) is bivariate discrete random variable with joint p. f. $p_{ij}, i,j=1,2,\cdots$. If $p_{+j}>0$ for given j, then

$$P\{X=x_i \mid Y=y_j\}=\frac{P\{X=x_i, Y=y_j\}}{P\{Y=y_j\}}=\frac{p_{ij}}{p_{+j}}, \quad i=1,2,\cdots \tag{3.3.1}$$

is the conditional p. f. of X given $Y=y_j$. Similarly, If $p_{i+}>0$ for given i, then

$$P\{Y=y_j \mid X=x_i\}=\frac{P\{X=x_i, Y=y_j\}}{P\{X=x_i\}}=\frac{p_{ij}}{p_{i+}}, \quad j=1,2,\cdots \tag{3.3.2}$$

is the conditional p. f. of Y given $X=x_i$.

Example 3.3.1

Consider a sequence of trials that a soldier shoots at a target. Suppose the trials are independent and the probability that the soldier hits the target is $p\in(0,1)$ each time. The soldier will stop shooting when he hits the target twice. Let X be the number of shooting when the soldier hits the target first, and Y be the total number of shooting. Find the joint p. f. of (X,Y) and the conditional p. f. s.

Solution.

It is known that $\{X=i\}$ is the event that the soldier firstly hits the target at the ith shooting, $\{Y=j\}$ is the event that the soldier secondly hits the target at the jth shooting, and $i<j$. However $\{X=i,Y=j\}$ is the event that the soldier only hits the target at the ith and jth shootings and does not hit the target at other times. Let $q=1-p$. We have

$$P\{X=i,Y=j\}=p^2q^{j-2},\quad i=1,2,\cdots,\quad j=i+1,i+2,\cdots.$$

The marginal p. f. s are

$$P\{X=i\}=\sum_{j=i+1}^{\infty}p^2q^{j-2}=pq^{i-1},\quad i=1,2,\cdots,$$

$$P\{Y=j\}=\sum_{i=1}^{j-1}p^2q^{j-2}=(j-1)p^2q^{j-2},\quad j=2,3,\cdots.$$

For given $j=2,3,\cdots$, the conditional p. f. of X given $Y=j$ is

$$P\{X=i|Y=j\}=\frac{p^2q^{j-2}}{(j-1)p^2q^{j-2}}=\frac{1}{j-1},\quad i=1,2,\cdots,j-1,$$

and for given $i=1,2,\cdots$, the conditional p. f. of Y given $X=i$ is

$$P\{Y=j|X=i\}=\frac{p^2q^{j-2}}{pq^{i-1}}=pq^{j-i-1},\quad j=i+1,i+2,\cdots.$$

3.3.2 Continuous Conditional Distribution

When (X,Y) is bivariate continuous random variable, $P\{X=x\}=P\{Y=y\}=0$ for any x and y. It is said that we can not define the continuous conditional distribution as discrete random variable. Next, we use the limit to deal with this continuous situation.

For given y and any $\varepsilon>0$, suppose that $P\{y-\varepsilon<Y\leqslant y+\varepsilon\}>0$, and for any x, we have

$$P\{X\leqslant x|y-\varepsilon<Y\leqslant y+\varepsilon\}=\frac{P\{X\leqslant x,y-\varepsilon<Y\leqslant y+\varepsilon\}}{P\{y-\varepsilon<Y\leqslant y+\varepsilon\}},$$

which is the condition d. f. of X given condition $y-\varepsilon<Y\leqslant y+\varepsilon$.

Definition 3.3.2

For given y and any $\varepsilon>0$, suppose that $P\{y-\varepsilon<Y\leqslant y+\varepsilon\}>0$. If, for any x, the limit

$$\lim_{\varepsilon\to 0^+} P\{X\leqslant x \mid y-\varepsilon<Y\leqslant y+\varepsilon\}=\lim_{\varepsilon\to 0^+}\frac{P\{X\leqslant x, y-\varepsilon<Y\leqslant y+\varepsilon\}}{P\{y-\varepsilon<Y\leqslant y+\varepsilon\}}$$

exists, then the limit is called as conditional d. f. of X given $Y=y$, denoted by $P\{X\leqslant x|Y=y\}$ or $F_{X|Y}(x|y)$. Similarly, the conditional d. f. of Y given the condition $X=x$ is defined as

$$\lim_{\varepsilon\to 0^+} P\{Y\leqslant y \mid x-\varepsilon<X\leqslant x+\varepsilon\},$$

if it exists, and denoted by $P\{Y\leqslant y|X=x\}$ or $F_{Y|X}(y|x)$.

Suppose that the bivariate random variable (X,Y) has the d. f. $F(x,y)$ and p. d. f. $f(x,y)$. If, at (x,y), the p. d. f. s $f(x,y)$ and $f_X(y)$ are continuous and $f_Y(y)>0$, then

$$\begin{aligned}F_{X|Y}(x|y)&=\lim_{\varepsilon\to 0^+}\frac{P\{X\leqslant x, y-\varepsilon<Y\leqslant y+\varepsilon\}}{P\{y-\varepsilon<Y\leqslant y+\varepsilon\}}\\&=\lim_{\varepsilon\to 0^+}\frac{F(x,y+\varepsilon)-F(x,y-\varepsilon)}{F_Y(y+\varepsilon)-F_Y(y-\varepsilon)}\\&=\lim_{\varepsilon\to 0^+}\frac{[F(x,y+\varepsilon)-F(x,y-\varepsilon)]/2\varepsilon}{[F_Y(y+\varepsilon)-F_Y(y-\varepsilon)]/2\varepsilon}\\&=\frac{\dfrac{\partial F(x,y)}{\partial y}}{\dfrac{\mathrm{d}F_Y(y)}{\mathrm{d}y}},\end{aligned}$$

that is,

$$F_{X|Y}(x \mid y)=\frac{\int_{-\infty}^{x} f(u,y)\,\mathrm{d}u}{f_Y(y)}=\int_{-\infty}^{x}\frac{f(u,y)}{f_Y(y)}\mathrm{d}u. \tag{3.3.3}$$

If we denote the conditional p. d. f. of X given condition $Y=y$ as $f_{X|Y}(x|y)$, then

$$f_{X|Y}(x|y)=\frac{\mathrm{d}}{\mathrm{d}x}F_{X|Y}(x|y)=\frac{f(x,y)}{f_Y(y)}. \tag{3.3.4}$$

Similarly, the conditional p. d. f. of Y given condition $X=x$ is

$$f_{Y|X}(y|x)=\frac{f(x,y)}{f_X(x)}. \tag{3.3.5}$$

The conditional p. d. f. $f_{X|Y}(x|y)$ satisfies that, for any given y,

$$f_{X|Y}(x \mid y) \geqslant 0 \quad \text{and} \quad \int_{-\infty}^{\infty} f_{X|Y}(x \mid y)\mathrm{d}x = 1.$$

Example 3. 3. 2

Suppose that (X, Y) has a uniform distribution on the region $D=\{(x,y): x^2+y^2 \leqslant 1\}$. Find the conditional p. d. f. $f_{X|Y}(x|y)$.

Solution.

The p. d. f. of (X,Y) is

$$f(x,y)=\begin{cases} \dfrac{1}{\pi}, & x^2+y^2 \leqslant 1; \\ 0, & \text{otherwise}, \end{cases}$$

and the marginal p. d. f. is

$$f_Y(y) = \int_{-\infty}^{\infty} f(x,y)\mathrm{d}x$$

$$= \begin{cases} \displaystyle\int_{-\sqrt{1-y^2}}^{+\sqrt{1-y^2}} \frac{1}{\pi}\mathrm{d}x = \frac{2}{\pi}\sqrt{1-y^2}, & -1 \leqslant y \leqslant 1; \\ 0, & \text{otherwise}. \end{cases}$$

So, for $-1<y<1$, the conditional p. d. f. of X given $Y=y$ is

$$f_{X|Y}(x|y)=\frac{f(x,y)}{f_Y(y)}$$

$$=\begin{cases} \dfrac{1/\pi}{\dfrac{2}{\pi}\sqrt{1-y^2}}=\dfrac{1}{2\sqrt{1-y^2}}, & -\sqrt{1-y^2} \leqslant x \leqslant \sqrt{1-y^2}; \\ 0, & \text{otherwise}. \end{cases}$$

Especially, if $y=0$, then

$$f_{X|Y}(x|y=0)=\begin{cases} \dfrac{1}{2}, & -1 \leqslant x \leqslant 1; \\ 0, & \text{otherwise}. \end{cases}$$

From formula (3. 3. 4) and formula (3. 3. 5), we get the multiplication formula

$$f(x,y)=f_X(x)f_{Y|X}(y|x)=f_Y(y)f_{X|Y}(x|y). \qquad (3.3.6)$$

Example 3. 3. 3

Suppose that $X \sim U(0,1)$. When $X=x \in (0,1)$, $Y \sim U(0,x)$. Find (1)

$P\{Y>X^2\}$; (2) the marginal p. d. f. $f_Y(y)$.

Solution.

Since $X \sim U(0,1)$, the p. d. f. of X is

$$f_X(x)=\begin{cases}1, & 0<x<1;\\ 0, & \text{otherwise.}\end{cases}$$

When $X=x\in(0,1)$, the conditional p. d. f. of Y is

$$f_{Y|X}(y|x)=\begin{cases}\frac{1}{x}, & 0<y<x;\\ 0, & \text{otherwise.}\end{cases}$$

By formula (3. 3. 6), the joint p. d. f. is

$$f(x,y)=f_X(x)f_{Y|X}(y|x)=\begin{cases}\frac{1}{x}, & 0<y<x<1;\\ 0, & \text{otherwise.}\end{cases}$$

(1) $P\{Y\geqslant X^2\}=\int_0^1\int_{x^2}^x\frac{1}{x}\mathrm{d}y\mathrm{d}x=\int_0^1(1-x)\mathrm{d}x=\frac{1}{2}$.

(2) $f_Y(y)=\int_{-\infty}^{\infty}f(x,y)\mathrm{d}x=\begin{cases}\int_y^1\frac{1}{x}\mathrm{d}x=-\ln y, & 0<y<1;\\ 0, & \text{otherwise.}\end{cases}$

3.4 Independence of Random Variables

We generalize the independence of events to random variables. Based on the independence of events, if a random variable takes any value, and it does not affect the probability that another random variable takes values, then the two random variables are independent intuitively. In this section, we define the independence for two and more random variables.

3.4.1 Independence for Two Random Variables

Definition 3.4.1 (Independence)

Suppose that X and Y are two given random variables, if for any x and

y, the events $\{X \leqslant x\}$ and $\{Y \leqslant y\}$ are independent, that is,

$$P\{X \leqslant x, Y \leqslant y\} = P\{X \leqslant x\} P\{Y \leqslant y\}, \tag{3.4.1}$$

then X and Y are independent.

Suppose that bivariate random variable (X, Y) has d. f. $F(x, y)$, marginal d. f. s $F_X(x)$ and $F_Y(y)$. Then formula (3.4.1) is equivalent to

$$F(x, y) = F_X(x) F_Y(y) \text{ for any } x, y. \tag{3.4.2}$$

By the definition of independence of X and Y, for any $x_1 < x_2$ and $y_1 < y_2$, the events $\{x_1 < X \leqslant x_2\}$ and $\{y_1 < Y \leqslant y_2\}$ are independent, because

$$\begin{aligned} & P\{x_1 < X \leqslant x_2, y_1 < Y \leqslant y_2\} \\ = & F(x_2, y_2) - F(x_1, y_2) - F(x_2, y_1) + F(x_1, y_1) \\ = & F_X(x_2) F_Y(y_2) - F_X(x_1) F_Y(y_2) - F_X(x_2) F_Y(y_1) + F_X(x_1) F_Y(y_1) \\ = & [F_X(x_2) - F_X(x_1)][F_Y(y_2) - F_Y(y_1)] \\ = & P\{x_1 < X \leqslant x_2\} P\{y_1 < Y \leqslant y_2\}. \end{aligned}$$

In fact, we have a more general result: for any interval I_1 and I_2, subsets of $\mathbf{R}$, the events $\{X \in I_1\}$ and $\{Y \in I_2\}$ are independent if X and Y are independent.

When (X, Y) is discrete or continuous, we have the following two results.

Theorem 3.4.1

Suppose that (X, Y) is discrete random variable with joint p. f.

$$P\{X = x_i, Y = y_j\} = p_{ij}, \quad i, j = 1, 2, \cdots$$

and marginal p. f. s

$$P\{X = x_i\} = p_{i+}, \quad i = 1, 2, \cdots;$$

$$P\{Y = y_j\} = p_{+j}, \quad j = 1, 2, \cdots.$$

Then X and Y are independent if and only if

$$p_{ij} = p_{i+} p_{+j}, \quad i, j = 1, 2, \cdots. \tag{3.4.3}$$

Proof.

We firstly prove the sufficiency. Suppose that $p_{ij} = p_{i+} p_{+j}, i, j = 1, 2, \cdots$. For any x, y,

$$F(x,y) = \sum_{x_i \leqslant x, y_j \leqslant y} p_{ij} = \sum_{x_i \leqslant x, y_j \leqslant y} p_{i+} p_{+j}$$
$$= \left(\sum_{x_i \leqslant x} p_{i\cdot}\right) \cdot \left(\sum_{y_j \leqslant y} p_{\cdot j}\right)$$
$$= F_X(x) F_Y(y).$$

So, X and Y are independent.

We secondly prove the necessity. Suppose that X and Y are independent, for any $x_1 < x_2$ and $y_1 < y_2$, we have

$$P\{x_1 < X \leqslant x_2, y_1 < Y \leqslant y_2\} = P\{x_1 < X \leqslant x_2\} P\{y_1 < Y \leqslant y_2\}.$$

For any i, j, by continuity of probability we have

$$p_{ij} = P\{X = x_i, Y = y_j\}$$
$$= \lim_{n,m \to \infty} P\{x_i - \frac{1}{n} < X \leqslant x_i, y_j - \frac{1}{m} < Y \leqslant y_j\}$$
$$= \lim_{n \to \infty} P\{x_i - \frac{1}{n} < X \leqslant x_i\} \cdot \lim_{m \to \infty} P\{y_j - \frac{1}{m} < Y \leqslant y_j\}$$
$$= P\{X = x_i\} P\{Y = y_j\} = p_{i+} p_{+j},$$

which means formula (3. 4. 3) holds.

Theorem 3. 4. 2

Suppose that (X, Y) is continuous random variable with joint p. d. f. $f(x,y)$ and marginal p. d. f. s $f_X(x)$ and $f_Y(y)$. Both X and Y are independent if and only if

$$f(x,y) = f_X(x) f_Y(y) \tag{3. 4. 4}$$

for all continuous points (x,y) of $f(x,y)$, $f_X(x)$ and $f_Y(y)$.

Proof.

We firstly prove the sufficiency. Suppose formula (3. 4. 4) holds, we have

$$F(x,y) = \int_{-\infty}^{x} \int_{-\infty}^{y} f(u,v) \mathrm{d}v \mathrm{d}u$$
$$= \int_{-\infty}^{x} \int_{-\infty}^{y} f_X(u) f_Y(v) \mathrm{d}v \mathrm{d}u$$
$$= \int_{-\infty}^{x} f_X(u) \mathrm{d}u \cdot \int_{-\infty}^{y} f_Y(v) \mathrm{d}v$$
$$= F_X(x) F_Y(y).$$

So X and Y are independent.

We secondly prove the necessity. Suppose that X and Y are independent, that is, $F(x,y)=F_X(x)F_Y(y)$ in formula (3.4.2) holds for any x, y. At the continuous point (x,y) of $f(x,y)$, $f_X(x)$ and $f_Y(y)$, we have

$$f(x,y)=\frac{\partial^2 F(x,y)}{\partial x \partial y}=\frac{\partial^2 F_X(x)F_Y(y)}{\partial x \partial y}$$

$$=\frac{\mathrm{d}F_X(x)}{\mathrm{d}x}\frac{\mathrm{d}F_Y(y)}{\mathrm{d}y}=f_X(x)f_Y(y),$$

which follows formula (3.4.4).

Example 3.4.1

Suppose that the joint p.f. of X and Y is specified by the following table 3.4.1.

Table 3.4.1

X \ Y	1	2	3	4	Total
1	0.06	0.02	0.04	0.08	0.20
2	0.15	0.05	0.10	0.20	0.50
3	0.09	0.03	0.06	0.12	0.30
Total	0.30	0.10	0.20	0.40	1

Since it can be found from this table that formula (3.4.3) is satisfied for all values of i and j. That is, $p_{ij}=p_{i+}p_{+j}$. It follows that X and Y are independent.

Example 3.4.2

Suppose that X and Y are independent random variables with identical p.d.f.:

$$f(x)=\begin{cases} 2e^{-2x}, & x>0; \\ 0, & \text{otherwise}. \end{cases}$$

Find the probability $P\{X+Y<1\}$.

Solution.

Since X and Y are independent, the joint p.d.f. is

$$f(x,y)=\begin{cases} 4e^{-2(x+y)}, & x>0, y>0; \\ 0, & \text{otherwise}. \end{cases}$$

So,

$$P\{X+Y<1\}=\iint\limits_{x+y<1} f(x,y)\,dxdy=\int_0^1 dx\int_0^{1-x} 4e^{-2(x+y)}\,dy=1-3e^{-2}.$$

Example 3.4.3

Suppose that $(X,Y)\sim N(\mu_1,\mu_2,\sigma_1^2,\sigma_2^2,\rho)$. Prove that X and Y are independent if and only if $\rho=0$.

Proof.

We firstly prove the sufficiency. If $\rho=0$, then for all x and y,

$$\begin{aligned} f(x,y) &= \frac{1}{2\pi\sigma_1\sigma_2}\exp\left\{-\frac{1}{2}\left[\left(\frac{x-\mu_1}{\sigma_1}\right)^2+\left(\frac{y-\mu_2}{\sigma_2}\right)^2\right]\right\} \\ &= \frac{1}{\sqrt{2\pi}\sigma_1}e^{-\frac{(x-\mu_1)^2}{2\sigma_1^2}}\frac{1}{\sqrt{2\pi}\sigma_2}e^{-\frac{(x-\mu_2)^2}{2\sigma_2^2}} \\ &= f_X(x)f_Y(y), \end{aligned}$$

which means that X and Y are independent.

We next prove the necessity. If X and Y are independent, then, for all x and y,

$$f(x,y)=f_X(x)\cdot f_Y(y).$$

Let $x=\mu_1$ and $y=\mu_2$ in the above equality, we have

$$f(\mu_1,\mu_2)=f_X(\mu_1)\cdot f_Y(\mu_2),$$

which follows

$$\frac{1}{2\pi\sigma_1\sigma_2\sqrt{1-\rho^2}}=\frac{1}{\sqrt{2\pi}\sigma_1}\cdot\frac{1}{\sqrt{2\pi}\sigma_2}.$$

So, $\rho=0$.

3.4.2 Independence for More than Two Random Variables

Definition 3.4.2

Suppose that $(X_1,X_2,\cdots,X_n)$ is a n-dimensional random vector. If for any given $x_1,x_2,\cdots,x_n$, the events $\{X_1\leqslant x_1\},\{X_2\leqslant x_2\},\cdots,\{X_n\leqslant x_n\}$ are independent, that is,

$$P\{X_1\leqslant x_1,X_2\leqslant x_2,\cdots,X_n\leqslant x_n\}=P\{X_1\leqslant x_1\}P\{X_2\leqslant x_2\}\cdots P\{X_n\leqslant x_n\}, \tag{3.4.5}$$

then $X_1, X_2, \cdots, X_n$ are independent.

If $X_1, X_2, \cdots, X_n$ has joint d. f. $F(x_1, x_2, \cdots, x_n)$ and marginal d. f. s $F_{X_1}(x_1), F_{X_2}(x_2), \cdots, F_{X_n}(x_n)$, $X_1, X_2, \cdots, X_n$ are independent if and only if

$$F(x_1, x_2, \cdots, x_n) = F_{X_1}(x_1) F_{X_2}(x_2) \cdots F_{X_n}(x_n), \tag{3.4.6}$$

because of formula (3.4.5). If $(X_1, X_2, \cdots, X_n)$ is a n-dimensional discrete random variable, formula (3.4.6) is equivalent to

$$P\{X_1 = x_1, X_2 = x_2, \cdots, X_n = x_n\} = P\{X_1 = x_1\} P\{X_2 = x_2\} \cdots P\{X_n = x_n\}$$

If $(X_1, X_2, \cdots, X_n)$ is a n-dimensional continuous random variable with p. d. f. $f(x_1, x_2, \cdots, x_n)$ and marginal p. d. f. s $f_{X_1}(x_1)$, $f_{X_2}(x_2)$, $\cdots$, $f_{X_n}(x_n)$, then formula (3.4.6) is equivalent to

$$f(x_1, x_2, \cdots, x_n) = f_{X_1}(x_1) f_{X_2}(x_2) \cdots f_{X_n}(x_n).$$

3.5 Functions of Two or More Random Variables

In this section, we discuss the distribution functions of two and more random variables.

3.5.1 Function of Discrete Random Variables

Here two examples are given for readers to get insights for distributions of functions of discrete random variables.

Example 3.5.1

Suppose that (X, Y) has the joint p. f. given in the following Table 3.5.1. Find the p. f. s of $X+Y$, $X-Y$ and $|XY|$.

Table 3.5.1

X \ Y	1	2
-1	$\frac{1}{4}$	$\frac{1}{4}$
1	$\frac{1}{3}$	$\frac{1}{6}$

Solution.

By the joint d. f. given, we have Table 3. 5. 2.

Table 3. 5. 2

P	$\frac{1}{4}$	$\frac{1}{4}$	$\frac{2}{6}$	$\frac{1}{6}$
(X,Y)	$(-1,1)$	$(-1,2)$	$(1,1)$	$(1,2)$
$X+Y$	0	1	2	3
$X-Y$	-2	-3	0	-1
$\|XY\|$	1	2	1	2

So, the p. f. s of $X+Y$, $X-Y$and $|XY|$ are Table 3. 5. 3, Table 3. 5. 4 and Table 3. 5. 5.

Table 3. 5. 3

$X+Y$	0	1	2	3
P	$\frac{1}{4}$	$\frac{1}{4}$	$\frac{2}{6}$	$\frac{1}{6}$

Table 3. 5. 4

$X-Y$	-2	-3	0	-1
P	$\frac{1}{4}$	$\frac{1}{4}$	$\frac{2}{6}$	$\frac{1}{6}$

Table 3. 5. 5

$\|XY\|$	1	2
P	$\frac{7}{12}$	$\frac{5}{12}$

Example 3. 5. 2

Suppose that X and Y are independent and identically distributed(i. i. d). random variables with distribution $B(1,p)$. Find the distribution of $X+Y$.

Solution.

Since X and Y has the same distribution $B(1,p)$, we have

$$P\{X=1\}=P\{Y=1\}=p=1-P\{X=0\}=1-P\{Y=0\}.$$

We have

$$P\{X+Y=0\}=P\{X=0,Y=0\}=P\{X=0\}P\{Y=0\}=(1-p)^2,$$

$$\begin{aligned}P\{X+Y=1\}&=P\{X=0,Y=1\}+P\{X=1,Y=0\}\\&=P\{X=0\}P\{Y=1\}+P\{X=1\}P\{Y=0\}=2p(1-p),\end{aligned}$$

$$P\{X+Y=2\}=P\{X=1,Y=1\}=P\{X=1\}P\{Y=1\}=p^2.$$

In summary, $X+Y\sim N(2,p)$.

Similarly with Example 3.5.2, if $X\sim B(n_1,p)$ and $Y\sim B(n_2,p)$, and they are independent, then $X+Y\sim B(n_1+n_2,p)$.

3.5.2 Function of Continuous Random Variables

Suppose that (X,Y) has a continuous d.f. $F(x,y)$ and p.d.f. $f(x,y)$. If $Z=g(X,Y)$ is a random variable, where $g(x,y)$ is a given continuous function, then Z is a continuous random variable generally. Next we discuss the p.d.f. of Z.

We assume that Z is a continuous random variable. Firstly, we find the d.f. of Z,

$$F_Z(z)=P\{Z\leqslant z\}=P\{g(X,Y)\leqslant z\}=\iint\limits_{g(x,y)\leqslant z} f(x,y)\mathrm{d}x\mathrm{d}y.$$

The p.d.f. of Z is $f_Z(z)=\frac{\mathrm{d}}{\mathrm{d}z}F_Z(z)$. However, the difficulty to compute $F_Z(z)$ is to determine the integral region $\{(x,y):g(x,y)\leqslant z\}$, which is easy for simple function, such as $g(x,y)=x+y$, and is difficult for general functions.

(1) The distribution of sum $Z=X+Y$.

Suppose that (X,Y) is a bivariate random variable with d.f. $F(x,y)$, $Z=X+Y$ is a continuous random variable, and its d.f. is, for $-\infty<z<+\infty$,

$$\begin{aligned}F_Z(z)&=P\{Z\leqslant z\}=P\{X+Y\leqslant z\}\\&=\iint\limits_{x+y\leqslant z} f(x,y)\mathrm{d}x\mathrm{d}y\\&=\int_{-\infty}^{+\infty}\left(\int_{-\infty}^{z-y} f(x,y)\mathrm{d}x\right)\mathrm{d}y,\end{aligned}$$

its p. d. f. is

$$f_Z(z) = \frac{\mathrm{d}}{\mathrm{d}z}F_Z(z)$$

$$= \int_{-\infty}^{+\infty}\left(\frac{\mathrm{d}}{\mathrm{d}z}\int_{-\infty}^{z-y} f(x,y)\mathrm{d}x\right)\mathrm{d}y$$

$$= \int_{-\infty}^{+\infty} f(z-y,y)\mathrm{d}y. \qquad (3.5.1)$$

Similarly, by symmetry we have

$$f_Z(z) = \int_{-\infty}^{+\infty} f(x,z-x)\mathrm{d}x, \quad -\infty < z < +\infty. \qquad (3.5.2)$$

If both X and Y are independent, and their p. d. f. s are $f_X(x)$ and $f_Y(y)$ respectively, then, formula (3.5.1) and formula (3.5.2) are equivalent to, for $-\infty<z<+\infty$,

$$f_Z(z) = \int_{-\infty}^{+\infty} f_X(x)f_Y(z-x)\mathrm{d}x = \int_{-\infty}^{+\infty} f_X(z-y)f_Y(y)\mathrm{d}y, \qquad (3.5.3)$$

which is called the convolution of the p. d. f. s f_X and f_Y, denoted by $f_X * f_Y$, that is,

$$f_Z(z)=f_X * f_Y(z).$$

Example 3.5.3

Suppose that (X,Y) has the p. d. f.

$$f(x,y)=\begin{cases}1, & 0<x<1, 0<y<2(1-x);\\ 0, & \text{otherwise.}\end{cases}$$

Find the p. d. f. of $Z=X+Y$.

Solution.

By formula (3.5.2), the p. d. f. of Z is

$$f_Z(z) = \int_{-\infty}^{+\infty} f(x,z-x)\mathrm{d}x, \quad -\infty < z < +\infty,$$

whose integrand $f(x,z-x)\neq 0$ if and only if

$$\begin{cases}0<x<1,\\ 0<z-x<2(1-x),\end{cases}$$

that is,

$$\begin{cases}0<x<1,\\ z>x,\\ z<2-x.\end{cases}$$

So, when $z\leqslant 0$ or $z\geqslant 2$, $f_Z(z)=0$; when $0<z<1$, $f_Z(z)=\int_0^z 1\mathrm{d}x=z$; when $1\leqslant z<2$, $f_Z(z)=\int_0^{2-z}1\,\mathrm{d}x=2-z$. Hence, the p. d. f. of Z is

$$f_Z(z)=\begin{cases}z, & 0<z<1;\\ 2-z, & 1\leqslant z<2;\\ 0, & \text{otherwise.}\end{cases}$$

Example 3.5.4

Suppose that X and Y are i. i. d. variable randoms with distribution $N(0,1)$, find the p. d. f. of $Z=X+Y$.

Solution.

Since X and Y are i. i. d. with distribution $N(0,1)$, we have

$$f_X(x)=\frac{1}{\sqrt{2\pi}}\mathrm{e}^{-\frac{x^2}{2}},\quad -\infty<x<+\infty;$$

$$f_Y(y)=\frac{1}{\sqrt{2\pi}}\mathrm{e}^{-\frac{y^2}{2}},\quad -\infty<y<+\infty.$$

By formula (3.5.3),

$$\begin{aligned}f_Z(z)&=\int_{-\infty}^{+\infty}f_X(x)f_Y(z-x)\mathrm{d}x\\&=\int_{-\infty}^{+\infty}\frac{1}{2\pi}\mathrm{e}^{-\frac{x^2}{2}}\cdot\mathrm{e}^{-\frac{(z-x)^2}{2}}\mathrm{d}x\\&=\frac{1}{\sqrt{2\pi}}\mathrm{e}^{-\frac{z^2}{4}}\cdot\frac{1}{\sqrt{2\pi}}\int_{-\infty}^{+\infty}\mathrm{e}^{-(x-\frac{z}{2})^2}\mathrm{d}x\\&=\frac{1}{\sqrt{2\pi}}\mathrm{e}^{-\frac{z^2}{4}}\cdot\sqrt{\frac{1}{2}}\left(\frac{1}{\sqrt{2\pi}\sqrt{\frac{1}{2}}}\int_{-\infty}^{+\infty}\mathrm{e}^{-\frac{(x-\frac{z}{2})^2}{2\times\frac{1}{2}}}\mathrm{d}x\right)\\&=\frac{1}{2\sqrt{\pi}}\mathrm{e}^{-\frac{z^2}{4}}.\end{aligned}$$

That is, $Z\sim N(0,2)$.

In general, if both X and Y are independent and $X \sim N(\mu_1, \sigma_1^2)$ and $Y \sim N(\mu_2, \sigma_2^2)$, then $X+Y \sim N(\mu_1+\mu_2, \sigma_1^2+\sigma_2^2)$. Furthermore, this result can be generalized to more general result: if $X_i \sim N(\mu_i, \sigma_i^2), i=1,2,\cdots,n$, and are independent, then the sum

$$\sum_{i=1}^{n} X_i \sim N(\sum_{i=1}^{n} \mu_i, \sum_{i=1}^{n} \sigma_i^2).$$

As a result, if we are given constants b and $a_i, i=1,2,\cdots,n$, then the sum

$$\sum_{i=1}^{n} a_i X_i + b \sim N(\sum_{i=1}^{n} a_i \mu_i + b, \sum_{i=1}^{n} a_i^2 \sigma_i^2),$$

where $a_i, i=1,2,\cdots,n$, are not all zeros.

(2) The distribution of functions $\max\{X_i, i=1,2,\cdots,n\}$ and $\min\{X_i, i=1,2,\cdots,n\}$.

Suppose that continuous random variables $X_1, X_2, \cdots, X_n$ are independent, and have p. d. f. s $f_{X_1}(x_1), f_{X_2}(x_2), \cdots, f_{X_n}(x_n)$ respectively. Let

$$M=\max\{X_i, i=1,2,\cdots,n\}, \quad N=\min\{X_i, i=1,2,\cdots,n\}.$$

We next find their d. f. s F_M and F_N. For any $-\infty < x < +\infty$,

$$\begin{aligned} F_M(x) &= P\{M \leqslant x\} \\ &= P\{\max\{X_i, i=1,2,\cdots,n\} \leqslant x\} \\ &= P\{X_1 \leqslant x, X_2 \leqslant x, \cdots, X_n \leqslant x\} \\ &= P\{X_1 \leqslant x\} P\{X_2 \leqslant x\} \cdots P\{X_n \leqslant x\} \\ &= F_{X_1}(x) F_{X_2}(x) \cdots F_{X_n}(x), \end{aligned}$$

and

$$\begin{aligned} F_N(x) &= P\{N \leqslant x\} \\ &= P\{\min\{X_i, i=1,2,\cdots,n\} \leqslant x\} \\ &= 1-P\{\min\{X_i, i=1,2,\cdots,n\} > x\} \\ &= 1-P\{X_1 > x, X_2 > x, \cdots, X_n > x\} \\ &= 1-P\{X_1 > x\} P\{X_2 > x\} \cdots P\{X_n > x\} \\ &= 1-(1-P\{X_1 \leqslant x\})(1-P\{X_2 \leqslant x\}) \cdots (1-P\{X_n \leqslant x\}) \\ &= 1-[1-F_{X_1}(x)][1-F_{X_2}(x)] \cdots [1-F_{X_n}(x)]. \end{aligned}$$

If random variables $X_1, X_2, \cdots, X_n$ are i. i. d. and have the identical d. f. $F(x)$, then

$$F_M(x)=[F(x)]^n, \quad F_N(x)=1-[1-F(x)]^n. \tag{3.5.4}$$

Example 3. 5. 5

Suppose that X and Y are i. i. d., and have the identical exponential distribution with the p. d. f.

$$f(x)=\begin{cases} e^{-x}, & x>0; \\ 0, & \text{otherwise.} \end{cases}$$

Find the p. d. f. s of $M=\max\{X,Y\}$ and $N=\min\{X,Y\}$.

Solution.

We first note that the d. f. s of X and Y are identical, and is specified as

$$F(x)=\begin{cases} 1-e^{-x}, & x>0; \\ 0, & \text{otherwise.} \end{cases}$$

By formula (3. 5. 4), for $y>0$, the p. d. f. of M is

$$f_M(y)=\frac{d}{dy}F_M(y)=\frac{d}{dy}[F^2(y)]=\frac{d}{dy}(1-e^{-y})^2=2e^{-y}(1-e^{-y}),$$

and for $z>0$, the p. d. f. of N is

$$f_N(z)=\frac{d}{dz}F_N(z)=\frac{d}{dz}\{1-[1-F(z)]\}^2=\frac{d}{dz}e^{-2z}=2e^{-2z}.$$

So, the p. d. f. s are

$$f_M(y)=\begin{cases} 2e^{-y}(1-e^{-y}), & y>0; \\ 0, & \text{otherwise,} \end{cases}$$

and

$$f_N(z)=\begin{cases} 2e^{-2z}, & z>0; \\ 0, & \text{otherwise.} \end{cases} \tag{3.5.5}$$

From Example formula (3. 5. 5), we find that N has an exponential distribution with parameter 2 given X and Y have the identical exponential distribution with parameter 1. In fact, we have more general result: Suppose that X_i have the identical exponential distribution with parameter $\lambda_i>0, i=1, 2, \cdots, n$, that is, the p. d. f. of X_i is

$$f_{X_i}(x)=\begin{cases}\lambda_i e^{-\lambda_i x}, & x>0;\\ 0, & \text{otherwise.}\end{cases}$$

If they are furthermore assumed to be independent, then $N=\min\{X_i, i=1,2,\cdots,n\}$ has p. d. f. specified as

$$f_N(z)=\begin{cases}\sum_{i=1}^{n}\lambda_i e^{-\sum_{i=1}^{n}\lambda_i z}, & z>0;\\ 0, & \text{otherwise.}\end{cases}$$

That is, N has an exponential distribution with parameter $\sum_{i=1}^{n}\lambda_i$.

Exercises

3.1 There are 10 blue balls and 2 red balls in a box, all balls are assumed to be the same except different color. Now someone takes out two balls from the box in different ways: (i) with replacement; (ii) without replacement. Let X and Y be the number of blue and red balls taker. Find the p. d. of X and Y in two different ways.

3.2 Tossing a fair coin 4 times, we let X be the numbers of heads and let Y be the absolute value of the minus of heads and tails. Find the p. d. of X and Y.

3.3 Three balls marked 1, 2, 3 in a box are taken at random twice with replacement. Find the joint p. f. of X and Y.

(1) X is the larger number on the balls both taken and Y is the sum of the numbers;

(2) X is the number on the first ball taken and Y is the larger number of the both taken.

3.4 Consider a sequence of independent Bernoulli trials, each of which is a success with probability $p>0$. Let X_1 be the number of failures preceding the first success, and let X_2 be the number of the failures between the first two successes. Find the joint probability function of X_1 and X_2.

3.5 Suppose that (X,Y) have the p. d. f.

$$f(x,y)=\begin{cases}k(6-x-y), & 0<x<2, 2<y<4;\\ 0, & \text{otherwise}\end{cases}$$

for some constant k. (1) Find k; (2) Compute $P\{X<1,Y<3\}$, $P\{X<1\}$, $P\{Y<3\}$; (3) Find the marginal p. d. f. s of X and Y, and the conditional p. d. f. s $f_{X|Y}(x|y)$ and $f_{Y|X}(y|x)$.

3.6 The joint p. d. f. of X and Y is given by

$$f(x,y)=\begin{cases} ke^{-(x+2y)}, & x>0, y>0; \\ 0, & \text{otherwise}. \end{cases}$$

(1) Compute ① $P\{X>1,Y<1\}$; ② $P\{X<Y\}$; ③ $P\{X<a\}$, a is a constant. (2) Find the marginal p. d. f. s of X and Y, and the conditional p. d. f. s $f_{X|Y}(x|y)$ and $f_{Y|X}(y|x)$. (3) Are X and Y independent?

3.7 The joint p. d. f. of X and Y is given by

$$f(x,y)=\begin{cases} c, & x^2+y^2\leqslant 1; \\ 0, & \text{otherwise} \end{cases}$$

for some c. (1) Determine c; (2) Find the marginal p. d. f. s of X and Y; (3) Compute the probability that D, the distance from the origin of the point selected is less than or equal to a, where a is a constant; (4) Find the marginal p. d. f. s of X and Y, and the conditional p. d. f. s $f_{X|Y}(x|y)$ and $f_{Y|X}(y|x)$.

3.8 Find the marginal p. f. s in Exercise 1.

3.9 Suppose that (X,Y) have a uniform distribution on the region $D=\{(x,y):0<x<1,|y|<x\}$, find the conditional p. d. f. s $f_{X|Y}(x|y)$ and $f_{Y|X}(y|x)$.

3.10 The joint p. d. f. of X and Y is given by

$$f(x,y)=\begin{cases} 2(x+y), & 0<x<y<1; \\ 0, & \text{otherwise}. \end{cases}$$

Determine (1) $P\{X<\frac{1}{2}\}$; (2) the marginal p. d. f. of X; (3) the conditional p. d. f. of Y given that $X=x$.

3.11 Suppose that X and Y are random variables. The marginal p. d. f. of X is

$$f_X(x)=\begin{cases} 3x^2, & 0<x<1; \\ 0, & \text{otherwise}. \end{cases}$$

The conditional p. d. f. of Y given that $X=x$ is

$$f_{Y|X}(y|x)=\begin{cases}\dfrac{3y^2}{x^3}, & 0<y<x;\\ 0, & \text{otherwise.}\end{cases}$$

Determine (1) the marginal p. d. f. of Y; (2) the conditional p. d. f. of X given that $Y=y$.

3.12 The joint p. d. f. of X and Y is given by

$$f(x,y)=\begin{cases}3e^{-3y}, & 0<x<1,\quad y>0;\\ 0, & \text{otherwise.}\end{cases}$$

Are X and Y independent? What if the joint p. d. f.

$$f(x,y)=\begin{cases}24xy, & 0<x<1, 0<y<1,\quad 0<x+y<1,\\ 0, & \text{otherwise.}\end{cases}$$

3.13 Let X and Y be independent and uniformly distributed over $(0,1)$. Compute $P\{X>Y^2\}$.

3.14 Suppose that X satisfies $P\{X=c\}=1$ for some constant c. Prove that X is independent of any random variable.

3.15 Suppose that X and Y are independent, $X\sim U(0,1)$, Y has a distribution with its p. d. f.

$$f(y)=\begin{cases}\dfrac{1}{2}e^{-y/2}, & y>0;\\ 0, & \text{otherwise.}\end{cases}$$

(1) Find the joint p. d. f. of X and Y; (2) compute the probability that the equation $a^2+2Xa+Y=0$ has real solutions.

3.16 Suppose that X and Y are independent, and their p. d. f. s are given

$$f_X(x)=\begin{cases}e^{-x}, & x>0;\\ 0, & \text{otherwise.}\end{cases}\qquad f_Y(y)=\begin{cases}2e^{-2y}, & y>0;\\ 0, & \text{otherwise.}\end{cases}$$

Let

$$Z=\begin{cases}0, & X\leqslant Y;\\ 1, & X>Y.\end{cases}$$

Find the p. f. and d. f. of random variable Z.

3.17 Suppose X and Y have the following p. f. s:

Exercise Table 3.17.1

X	0	1	2
P	$\frac{1}{2}$	$\frac{1}{4}$	$\frac{1}{4}$

and

Exercise Table 3.17.2

Y	0	1
P	$\frac{1}{3}$	$\frac{2}{3}$

Find (1) the joint p. f. of X and Y; (2) the p. f. s of $Z_1 = X - Y$ and $Z_2 = X + Y$.

3.18 Suppose that X and Y are independent binomial random variables with different parameters, that is, $X \sim B(n_1, p)$, $Y \sim B(n_2, p)$ for some positive $p \in (0,1)$ and positive integer n_1 and n_2. Prove that $X + Y \sim B(n_1 + n_2, p)$.

3.19 Suppose that X and Y are i. i. d. geometric random variables, the p. f. is

$$P\{X = k\} = (1-p)^{k-1} p, \quad k = 1, 2, \cdots.$$

Let $Z = X + Y$. Find the p. f. of Z.

3.20 Suppose that X and Y are independent Poisson random variables with parameters $\lambda_1 > 0$ and $\lambda_2 > 0$ respectively. Let $Z = X + Y$.

(1) Prove that Z is a Poisson random variable with parameter $\lambda_1 + \lambda_2$;

(2) Generalize the above statement to the case with n random variables. That is, if X_i are independent Poisson random variables with parameters $\lambda_i > 0$ respectively, $i = 1, 2, \cdots, n$, then $Z \equiv X_1 + X_2 + \cdots + X_n$ is a Poisson random variable with parameter $\lambda_1 + \lambda_2 + \cdots + \lambda_n$.

3.21 Suppose that X and Y are independent and uniformly distributed over $[0,1]$. Let $Z = X + Y$. Find the p. d. f. s of Z and $Z/2$.

3.22 The joint p. d. f. of X and Y is given by

$$f(x,y)=\begin{cases}2(x+y), & 0<x<y<1;\\ 0, & \text{otherwise}.\end{cases}$$

Let $Z=X+Y$. Find the p. d. f. of Z.

3.23 Suppose that X and Y are i. i. d. random variables with p. d. f. as

$$f(x)=\begin{cases}e^{-x}, & x>0;\\ 0, & \text{otherwise}.\end{cases}$$

Find the p. d. f. s of $X+Y, X-Y$ and X/Y.

3.24 Suppose that X and Y are independent random variables, $X\sim U(0,1)$ and Y has a exponential distribution with parameter 1. Find the p. d. f. of $Z=X+Y$.

3.25 Suppose that X and Y have the following joint p. d. f.

$$f(x,y)=\begin{cases}\frac{1}{2}(x+y)e^{-(x+y)}, & x>0, y>0;\\ 0, & \text{otherwise}.\end{cases}$$

(1) Are X and Y independent? (2) Find the p. d. f. of $Z=X+Y$.

3.26 Suppose that X and Y have the following joint p. d. f.

$$f(x,y)=\frac{1}{\pi(1+x^2+y^2)^2}.$$

Find the p. d. f. of $Z=\sqrt{X^2+Y^2}$.

3.27 Suppose that X and Y are i. i. d. normal random variables with distribution $N(0,\sigma^2)$. Let $Z=\sqrt{X^2+Y^2}$. Verify the p. d. f. of Z is

$$f(z)=\begin{cases}\frac{z}{\sigma^2}e^{-z^2/(2\sigma^2)}, & z>0;\\ 0, & \text{otherwise}.\end{cases}$$

Here we note that Z is said to have a Rayleigh distribution.

3.28 Suppose that X and Y have the following joint p. d. f.

$$f(x,y)=\begin{cases}ce^{-(x+y)}, & 0<x<\infty, 0<y<\infty;\\ 0, & \text{otherwise}\end{cases}$$

for some constant c. (1) Determine c; (2) Find the marginal p. d. f. of X and Y; (3) Find the p. d. f. of $U=\max\{X,Y\}$ and $V=\min\{X,Y\}$.

3.29 Let $X_1, X_2, \cdots, X_n$ be i. i. d. over $(0,1)$. Find the p. d. f. s of $M=$

$\min\{X_1, x_2, \cdots, X_n\}$, $N=\max\{X_1, x_2, \cdots, X_n\}$.

3.30 Let $X_1, X_2, \cdots, X_n$ be i. i. d. normal random variables $N(0,1)$. Let $U=X+Y$, $V=X-Y$. Find the joint p. d. f. of U and V.

3.31 Suppose that X and Y have a joint p. d. f. $f(x,y)$. Find general formulas of p. d. f. s for random variables $X-Y$, XY and X/Y.

Chapter 4 Expectation

The distribution of a random variable X contains all of the statistical information about X. The distribution of X, however, is usually too cumbersome for presenting this information. For example, random variable X denotes the weight of any male student in a given high school, we need some specific information to know the physical development, such that, the expected weight and the deviation around it. Suppose that another random variable Y is given to denote the height of any male student in that high school. On the one hand, we need the expected height and the deviation around it too. On the other hand, we want to understand the relationship between X and Y.

In this chapter, we will study the expectation and variance of one random variable, covariance of two or more random variables, which represent the expected value, deviation and the correlation mentioned in the last paragraph.

4.1 Expectation of Random Variable

In this section, we firstly study the expectations of a random variable and function of random variables, and then properties of expectation.

4.1.1 the Concept of Expectation

We first see an example.

Example 4.1.1

Suppose that a soldier shoots at a target 100 times, the shooting records are in the following Table 4.1.1.

Table 4.1.1

Shooting rings k	0	1	2	3	4	5
Number of hittings n_k	5	15	20	10	20	30
Frequency of hittings $\frac{n_k}{n}$	$\frac{5}{100}$	$\frac{15}{100}$	$\frac{20}{100}$	$\frac{10}{100}$	$\frac{20}{100}$	$\frac{30}{100}$

Solution.

The average number of hittings

$$=\frac{\text{The total number of hitting rings}}{\text{The total number of shooting rings}}$$

$$=\frac{0\times5+1\times15+2\times20+3\times10+4\times20+5\times30}{100}$$

$$=0\times\frac{5}{100}+1\times\frac{15}{100}+2\times\frac{20}{100}+3\times\frac{10}{100}+4\times\frac{20}{100}+5\times\frac{30}{100}$$

$$=\sum_{k=0}^{5}k\times\frac{n_k}{n}$$

$$=3.15.$$

That is, the average rings are 3.15.

We first give the definition of expectation for a discrete random variable.

Definition 4.1.1

Suppose that a random variable X has a discrete distribution for which the p. f. is $P\{X=x_i\}=p_i, i=1,2,\cdots$. The expectation (or mean) of X, denoted by $E(X)$, is defined as

$$E(X)=\sum_{i=1}^{\infty}x_ip_i, \tag{4.1.1}$$

if

$$\sum_{i=1}^{\infty} | x_i | p_i < +\infty,$$

and does not exist otherwise.

Remark 4.1.1

There is a discrete X whose expectation does not exist. For example, suppose that X is a discrete random variable with p. f.

Table 4.1.2

X	2	$\frac{2^2}{2}$	$\frac{2^3}{3}$	…	$\frac{2^k}{k}$	…
P	$\frac{1}{2}$	$\frac{1}{4}$	$\frac{1}{8}$	…	$\frac{1}{2^k}$	…

Since

$$\sum_{i=1}^{\infty} x_i p_i = \sum_{i=1}^{\infty} \frac{2^k}{k} \cdot \frac{1}{2^k} = \sum_{k=1}^{\infty} \frac{1}{k} = +\infty,$$

it says that the expectation does not exist.

Remark 4.1.2

The expectation of X depends only on the distribution of X. Although $E(X)$ is called the expectation of X, it depends only on the distribution of X. Every two random variables that have the same distribution will have the same expectation even if they have nothing to do with each other. For this reason, we shall often refer to the expectation of a distribution even if we do not have in mind a random variable with that distribution.

Example 4.1.2

Suppose that a random variable X can have only the four different values $-2, 0, 1$ and 4, and that $P\{X=-2\}=0.1$, $P\{X=0\}=0.4$, $P\{X=1\}=0.3$, and $P\{X=4\}=0.2$. Then

$$E(X) = -2\times 0.1 + 0\times 0.4 + 1\times 0.3 + 4\times 0.2 = 0.9.$$

Example 4.1.3

Suppose that a random variable $X \sim B(1, p)$ with $p \in (0,1)$. Then

$$E(X)=0\cdot P\{X=0\}+1\cdot P\{X=1\}=p.$$

Example 4.1.4

Suppose that a random variable has a Poisson distribution with parameter $\lambda>0$, that is $X\sim P(\lambda)$. Then

$$E(X)=\sum_{k=0}^{\infty}kP\{X=k\}=\sum_{k=0}^{\infty}k\frac{\lambda^k}{k!}e^{-\lambda}=\lambda e^{-\lambda}\sum_{k=1}^{\infty}\frac{\lambda^{k-1}}{(k-1)!}=\lambda.$$

We define the expectation of a continuous random variable below.

Definition 4.1.2

Suppose that a random variable X has a continuous distribution with p. d. f. f. The expectation (or mean) of X, denoted by $E(X)$, is defined as

$$E(X)=\int_{-\infty}^{+\infty}xf(x)dx, \tag{4.1.2}$$

if

$$\int_{-\infty}^{+\infty}|x|f(x)dx<\infty,$$

and does not exist otherwise.

Remark 4.1.3

There is a continuous X whose expectation does not exist. For example, suppose that a random variable X has a continuous distribution for which the p. d. f. is as follows:

$$f(x)=\frac{1}{\pi(1+x^2)}, \quad -\infty<x<\infty. \tag{4.1.3}$$

This distribution is called the Cauchy distribution. Since

$$\int_{-\infty}^{+\infty}|x|f(x)dx=\frac{2}{\pi}\int_{0}^{+\infty}\frac{x}{1+x^2}dx=+\infty,$$

the expectation does not exist.

Example 4.1.5

Suppose that the p. d. f. of a random variable X with a continuous distribution

$$f(x)=\begin{cases}2x, & 0<x<1;\\ 0, & \text{otherwise.}\end{cases}$$

Then

$$E(X)=\int_{-\infty}^{+\infty}xf(x)\mathrm{d}x=\int_0^1 2x^2\mathrm{d}x=\frac{2}{3}.$$

Now we compute three special random variables.

Example 4.1.6

Suppose $X\sim U(a,b)$ with constants $a<b$. Then

$$E(X)=\int_{-\infty}^{+\infty}xf(x)\mathrm{d}x=\int_a^b\frac{x}{b-a}\mathrm{d}x=\frac{a+b}{2}.$$

Example 4.1.7

Suppose the continuous random variable X has a exponential distribution with parameter $\lambda>0$, that is, the p. d. f. is

$$f(x)=\begin{cases}\lambda \mathrm{e}^{-\lambda x}, & x>0;\\ 0, & \text{otherwise.}\end{cases}$$

Then

$$E(X)=\int_{-\infty}^{+\infty}xf(x)\mathrm{d}x=\int_0^{+\infty}\lambda x\mathrm{e}^{-\lambda x}\mathrm{d}x=\frac{1}{\lambda}.$$

Example 4.1.8

Suppose the continuous random variable X has a normal distribution $N(\mu,\sigma^2)$ for constants μ and σ^2, that is, the p. d. f. is

$$f(x)=\frac{1}{\sqrt{2\pi}\sigma}\mathrm{e}^{-\frac{(x-\mu)^2}{2\sigma^2}},\quad -\infty<x<+\infty.$$

Then

$$E(X)=\int_{-\infty}^{+\infty}xf(x)\mathrm{d}x=\int_{-\infty}^{+\infty}x\frac{1}{\sqrt{2\pi}\sigma}\mathrm{e}^{-\frac{(x-\mu)^2}{2\sigma^2}}\mathrm{d}x.$$

Define the integral transformation

$$t=\frac{x-\mu}{\sigma},$$

we have

$$\begin{aligned}E(X)&=\int_{-\infty}^{+\infty}xf(x)\mathrm{d}x=\frac{1}{\sqrt{2\pi}}\int_{-\infty}^{+\infty}(\sigma t+\mu)\mathrm{e}^{-\frac{t^2}{2}}\mathrm{d}t\\&=\frac{1}{\sqrt{2\pi}}\int_{-\infty}^{+\infty}\sigma t\mathrm{e}^{-\frac{t^2}{2}}\mathrm{d}t+\frac{\mu}{\sqrt{2\pi}}\int_{-\infty}^{+\infty}\mathrm{e}^{-\frac{t^2}{2}}\mathrm{d}t\\&=\mu.\end{aligned}$$

4.1.2 the Expectation of Function of Random Variables

Suppose that X is a random variable with d. f. F. Let g is a function, what is the expectation $E(g(X))$ of random variable $g(X)$? In general, if the distribution of $g(X)$ is given, we can compute the expectation by formula (4.1.1) and formula (4.1.2). However this computation method is too complex for us because we need first to find the distribution of $g(X)$. We next summarize these results in the following theorem.

Theorem 4.1.1

Suppose that X is a random variable and g is a continuous function. Let $Y=g(X)$.

(1) If X is a discrete random variable with p. f.

$$P\{X=x_i\}=p_i, \quad i=1,2,\cdots,$$

and $\sum_{i=1}^{\infty} g(x_i)p_i$ is absolute convergent, then

$$E(Y)=E(g(X))=\sum_{i=1}^{\infty} g(x_i)p_i. \tag{4.1.4}$$

(2) If X is a continuous random variable with p. d. f. $f_X(x)$ and

$$\int_{-\infty}^{+\infty} |g(x)| f_X(x)\mathrm{d}x<+\infty,$$

then

$$E(Y)=E(g(X))=\int_{-\infty}^{+\infty} g(x)f_X(x)\mathrm{d}x. \tag{4.1.5}$$

Proof.

We only give proofs for two special cases.

(1) Suppose that the distribution of X is discrete. Then the distribution of Y must also be discrete. For this case,

$$\begin{aligned} E(g(X)) &= \sum_j y_j P\{g(X)=y_j\} \\ &= \sum_j y_j \sum_{i:g(x_i)=y_j} p_i \\ &= \sum_j \sum_{i:g(x_i)=y_j} g(x_i)p_i \\ &= \sum_i g(x_i)p_i = E(Y). \end{aligned}$$

(2) Suppose that the distribution of X is continuous, then $Y=g(x)$ is a continuous random variable with p. d. f. f_Y. Suppose also that $g(x)$ is either strictly increasing or strictly decreasing, and that the inverse function $h(y)$ can be differentiated. Then, if we change variables from x to $y=g(x)$,

$$\begin{aligned} E(g(X)) &= \int_{-\infty}^{+\infty} g(x) f_X(x) \mathrm{d}x \\ &= \int_{-\infty}^{+\infty} y f_X[h(y)] \left| \frac{\mathrm{d}h(y)}{\mathrm{d}y} \right| \mathrm{d}y \\ &= \int_{-\infty}^{+\infty} y f_Y(y) \mathrm{d}y \\ &= E(Y). \end{aligned}$$

Example 4.1.9

Suppose that the p. d. f. of X is as given in Example 4.1.5 and that $Y=X^{1/2}$. Then, by formula (4.1.5)

$$E(Y) = \int_{-\infty}^{+\infty} x^{1/2} f(x) \mathrm{d}x = \int_0^1 x^{1/2}(2x) \mathrm{d}x = \frac{4}{5}.$$

Next, we generalize Theorem 4.1.1. to the case of functions of two random variables in the following theorem, whose proof is omitted.

Theorem 4.1.2

Suppose that (X,Y) is a bivariate random variable and $g(x,y)$ is a binary function. Let $Z=g(X,Y)$.

(1) If (X,Y) is a bivariate discrete random variable with p. f.

$$P\{X=x_i, Y=y_j\}=p_{ij}, \quad i,j=1,2,\cdots,$$

and $\sum_{i=1}^{\infty}\sum_{j=1}^{\infty} |g(x_i,y_j)| p_{ij} < \infty$, then

$$E(Z) = E[g(X,Y)] = \sum_{i=1}^{\infty}\sum_{j=1}^{\infty} g(x_i,y_j) p_{ij}. \tag{4.1.6}$$

(2) If (X,Y) is a bivariate continuous random variable with p. d. f. $f(x,y)$ and

$$\int_{-\infty}^{+\infty}\int_{-\infty}^{+\infty} |g(x,y)| f(x,y) \mathrm{d}x\mathrm{d}y < \infty,$$

then

$$E(Z) = E(g(X,Y)) = \int_{-\infty}^{+\infty}\int_{-\infty}^{+\infty} g(x,y) f(x,y) \mathrm{d}x\mathrm{d}y. \tag{4.1.7}$$

Example 4.1.10

Suppose that a point (X,Y) is chosen at random from the square $S=\{(x,y):0<x<1,0<y<1\}$. Determine the expected value of X^2+Y^2.

Solution

Since (X,Y) have a uniform distribution over the square S, and since the area size of S is 1, the joint p.d.f. of X and Y is

$$f(x,y)=\begin{cases}1, & (x,y)\in S,\\ 0, & \text{otherwise.}\end{cases}$$

Therefore,

$$\begin{aligned}E(X^2+Y^2) &= \int_{-\infty}^{+\infty}\int_{-\infty}^{+\infty}(x^2+y^2)f(x,y)\mathrm{d}x\mathrm{d}y \\ &= \int_0^1\int_0^1(x^2+y^2)\mathrm{d}x\mathrm{d}y \\ &= \frac{2}{3}.\end{aligned}$$

4.1.3 Properties of the expectation

Suppose that X is a random variable for which the expectation $E(X)$ exists. We shall present several results pertaining to the basic properties of expectations.

Proposition 4.1.1

If $Y=aX+b$, where a and b are constants, then

$$E(Y)=aE(X)+b.$$

Proof.

First we shall assume, for convenience, that X has a continuous distribution for which the p.d.f. is f. Then

$$\begin{aligned}E(Y) &= E(aX+b) \\ &= \int_{-\infty}^{+\infty}(ax+b)f(x)\mathrm{d}x \\ &= a\int_{-\infty}^{+\infty}xf(x)\mathrm{d}x+b\int_{-\infty}^{+\infty}f(x)\mathrm{d}x \\ &= aE(X)+b.\end{aligned}$$

Proposition 4.1.2

If X_1, X_2 are 2 random variables such that each expectation $E(X_i)$ exists, $i=1,2$, then

$$E(X_1+X_2)=E(X_1)+E(X_2). \tag{4.1.8}$$

Proof.

We assume that X_1 and X_2 have a continuous joint distribution for which the joint p. d. f. is $f(x_1, x_2)$, and the marginal p. d. f. s are $f_{X_1}(x_1)$ and $f_{X_2}(x_2)$ respectively. Then

$$\begin{aligned}
E(X_1+X_2) &= \int_{-\infty}^{+\infty}\int_{-\infty}^{+\infty}(x_1+x_2)f(x_1,x_2)\mathrm{d}x_1\mathrm{d}x_2 \\
&= \int_{-\infty}^{+\infty}\int_{-\infty}^{+\infty}x_1 f(x_1,x_2)\mathrm{d}x_1\mathrm{d}x_2 + \int_{-\infty}^{+\infty}\int_{-\infty}^{+\infty}x_2 f(x_1,x_2)\mathrm{d}x_1\mathrm{d}x_2 \\
&= \int_{-\infty}^{+\infty}x_1 f_{X_1}(x_1)\mathrm{d}x_1 + \int_{-\infty}^{+\infty}x_2 f_{X_2}(x_2)\mathrm{d}x_2 \\
&= E(X_1)+E(X_2).
\end{aligned}$$

The proof for a discrete distribution or a more general joint distribution is similar.

Remark 4.1.4

We note, the formula (4.1.8) can be generalized to the case of n random variables: If $X_1, \cdots, X_n$ are n random variables such that each expectation $E(X_i)$ exists, $i=1,\cdots,n$, then

$$E(X_1+\cdots+X_n)=E(X_1)+\cdots+E(X_n). \tag{4.1.9}$$

Example 4.1.11

Suppose that X has a binomial distribution, that is, $X\sim B(n,p)$ with $p\in(0,1)$. Find $E(X)$.

Solution.

Suppose $X_i\sim B(1,p)$, $i=1,2,\cdots,n$ and they are independent. Then

$$X_1+X_2+\cdots+X_n\sim B(n,p).$$

Since $X\sim B(n,p)$, we have

$$E(X)=E\Big(\sum_{i=1}^{n}X_i\Big)=\sum_{i=1}^{n}E(X_i)=np.$$

Example 4. 1. 12

Suppose that a person types n letters and n envelopes, and then places n letters into the n envelopes in a random manner with that one envelop is put one letter only. Let X be the number of letters that are placed in the correct envelopes. Find the mean of X.

Solution.

For $i=1, \cdots, n$, let $X_i=1$ if the ith letter is placed in the correct envelope, and let $X_i=0$ otherwise. Then, for $i=1,\cdots,n$,

$$P\{X_i=1\}=\frac{1}{n} \text{ and } P\{X_i=0\}=1-\frac{1}{n}.$$

Therefore,

$$E(X_i)=\frac{1}{n}, \quad i=1,\cdots,n.$$

Since $X=X_1+\cdots+X_n$, it follows that

$$\begin{aligned}E(X)&=E(X_1)+\cdots+E(X_n)\\&=\frac{1}{n}+\cdots+\frac{1}{n}\\&=1.\end{aligned}$$

Remark 4. 1. 5

From Example 4. 1. 12, the expected value of the number of correct matches of letters and envelopes is 1, regardless of the value of n.

Proposition 4. 1. 3

If X_1, X_2 are 2 independent random variables such that each expectation $E(X_i)$ exists, $i=1,2$, then

$$E(X_1X_2)=E(X_1)E(X_2). \tag{4.1.10}$$

Proof.

Suppose that X_1, X_2 have joint p. d. f. $f(x_1, x_2)$ and marginal p. d. f. s $f_{X_1}(x_1)$, $f_{X_2}(x_2)$. Since X_1, X_2 are independent, we have $f(x_1, x_2)=f_{X_1}(x_1)f_{X_2}(x_2)$.

Therefore,

$$
\begin{aligned}
E(X_1X_2) &= \int_{-\infty}^{+\infty}\int_{-\infty}^{+\infty} x_1x_2 f(x_1,x_2)\mathrm{d}x_1\mathrm{d}x_2 \\
&= \int_{-\infty}^{+\infty}\int_{-\infty}^{+\infty} x_1x_2 f_{X_1}(x_1)f_{X_2}(x_2)\mathrm{d}x_1\mathrm{d}x_2 \\
&= \int_{-\infty}^{+\infty} x_1 f_{X_1}(x_1)\mathrm{d}x_1 \cdot \int_{-\infty}^{+\infty} x_2 f_{X_2}(x_2)\mathrm{d}x_2 \\
&= E(X_1)E(X_2).
\end{aligned}
$$

The proof for a discrete distribution or a more general type of distribution is similar.

Remark 4.1.6

We note, the formula (4.1.10) can be generalized to the case of n independent random variables: If $X_1, \cdots, X_n$ are n independent random variables such that each expectation $E(X_i)$ exists $(i=1,\cdots,n)$, then

$$E(X_1\cdots X_n)=E(X_1)\cdots E(X_n). \tag{4.1.11}$$

Example 4.1.13

Suppose that X_1, X_2 and X_3 are independent random variables such that $E(X_i)=0$ and $E(X_i^2)=1$ for $i=1,2,3$. Determine the value of $E((X_1-1)^2(X_2-4X_3)^2)$.

Solution.

Since X_1, X_2 and X_3 are independent, it follows that two random variables $(X_1-1)^2$ and $(X_2-4X_3)^2$ are also independent. Therefore,

$$
\begin{aligned}
&E((X_1-1)^2(X_2-4X_3)^2) \\
&=E((X_1-1)^2)E((X_2-4X_3)^2) \\
&=E(X_1^2-2X_1+1)E(X_2^2-8X_2X_3+16X_3^2) \\
&=[E(X_1^2)-2E(X_1)+1][E(X_2^2)-8E(X_2X_3)+16E(X_3^2)] \\
&=2[1-8E(X_2)E(X_3)+16] \\
&=34.
\end{aligned}
$$

Proposition 4.1.4

If there exists a constant a such that $P\{X\geqslant a\}=1$, then $E(X)\geqslant a$. If there exists a constant b such that $P\{X\leqslant b\}=1$, then $E(X)\leqslant b$.

Proof.

For convenience, we assume that X has a continuous distribution with the p. d. f. f, and we shall suppose first that $P\{X \geqslant a\}=1$. Then

$$E(X)=\int_{-\infty}^{\infty} x f(x) \mathrm{d} x=\int_{a}^{\infty} x f(x) \mathrm{d} x \geqslant \int_{a}^{\infty} a f(x) \mathrm{d} x=a P\{X \geqslant a\}=a.$$

The proof of the other part of the theorem and the proof for a discrete distribution or a more general type of distribution are similar.

It follows from Propositon 4. 1. 4. that if $P\{a \leqslant X \leqslant b\}=1$ and $E(x)$ cxists, then $a \leqslant E(X) \leqslant b$.

We next prove the Markov inequality, which is used to prove the following proposition.

Lemma 4. 1. 1 (The Markov Inequality)

Suppose that X is a random variable such that $P\{X \geqslant 0\}=1$ and $E(x)$ exists. Then for every given number $t>0$,

$$P\{X \geqslant t\} \leqslant \frac{E(X)}{t}. \tag{4.1.12}$$

Proof.

For any $t>0$, we say

$$X \geqslant t \quad \mathbf{1}(x \geqslant t)$$

where $\mathbf{1}_A(x)=1$ if $x \in A$ and $\mathbf{1}_A(x)=0$ if $x \notin A$. Then by Proposition 4. 1. 5,

$$\begin{aligned} E(X) & \geqslant E(t\, \mathbf{1}(x \geqslant t)) \\ & =t\, E(\mathbf{1}(x \geqslant t)) \\ & =t\, P(x \geqslant t) \end{aligned}$$

Since $t>0$, We have

$$P(x \geqslant t) \leqslant \frac{E(X)}{t}$$

The Markov inequality is maily of interest for large values of t. In fact, when $t \leqslant E(X)$, the inequality is of no interest, because it is known that $P\{X \geqslant t\} \leqslant 1$. However, it can be found from the Markov inequality that for every nonnegative random variable X whose mean is 1, the maximum possible value of $P\{X \geqslant 100\}$ is 0. 01. Furthermore, it can be verified that this maximum value is actually attained by the random variable X satisfying $P\{X=0\}=0.99$ and $P\{X=100\}=0.01$.

Proposition 4. 1. 5

Suppose that the random variable X satisfies $P\{X \geqslant a\}=1$ and if $E(X)=a$.

Then $P\{X=a\}=1$.

Proof.

Suppose that $P\{X=a\}<1$. We assume that there exists a position constant $\varepsilon>0$ such that $P\{X-a\geqslant\varepsilon\}>0$. By Markov inequality,

$$P\{X-a\geqslant\varepsilon\}\leqslant\frac{E(X-a)}{\varepsilon}=0,$$

which contradicts to the hypothesis. Hence the result holds.

4.2 Variance and Moments

Although the mean of a distribution is a useful summary, it does not convey very much information about the distribution. For example, a random variable X with expectation 1 has the same expectation as the constant Y such that $P\{Y=1\}=1$ even if X is not constant. To distinguish the distribution of X from the distribution of Y in this case, it might be useful to give some measure of how spread out the distribution of X is. The variance of X is one such measure. In addition, we also need the expectations of powers X^k (called moments) for $k>2$ to understand its distribution. In this section, we introduce variance and moments of a distribution.

4.2.1 Variance

Suppose that X is a random variable with expectation μ and $P\{X=\mu\}\neq1$. We have three ways to measure the deviation from the expectation μ. (1) The first one is $X-\mu$ satisfies that $E(X-\mu)=0$, which makes no sense if we measure the expected deviation; (2) The second one is $|X-E(X)|$ which we use to define Chebyshev inequality; (3) The third one is $(X-E(X))^2$ which we use to define the variance.

Definition 4.2.1

Suppose that X is a random variable with mean $\mu=E(X)$. The variance of X, denoted by $\mathrm{Var}(X)$, is defined as

$$\mathrm{Var}(X)=E((X-\mu)^2), \tag{4.2.1}$$

if it exists. The standard deviation of X is defined to be as the square-root of the variance $\sqrt{\mathrm{Var}(X)}$.

Remark 4.2.1

(1) Firstly, we note that the variance, as well as the standard deviation, of a random variable X depends only on the distribution of X, just as the expectation of X only depends on the distribution.

(2) Since $\mathrm{Var}(X)$ is the expected value of the nonnegative random variable $(X-\mu)^2$, it follows that $\mathrm{Var}(X)\geqslant 0$. If the expectation in formula (4.2.1) is infinite, then $\mathrm{Var}(X)$ does not exist. However, if the possible values of X are bounded, then $\mathrm{Var}(X)$ must exist.

Suppose that X is a discrete random variable with expectation μ and p. f.

$$P\{X=x_i\}=p_i, \quad i=1,2,\cdots,$$

then, formula (4.2.1) is equivalent to

$$\mathrm{Var}(X) = \sum_{i=1}^{\infty}(x_i-\mu)^2 p_i.$$

If X is a continuous random variable with expectation μ and p. d. f. $f(x)$, then, formula (4.2.1) is equivalent to

$$\mathrm{Var}(X) = \int_{-\infty}^{+\infty}(x-\mu)^2 f(x)\mathrm{d}x.$$

We often use the following formula to compute the variance of X,

$$\mathrm{Var}(X)=E(X^2)-(E(X))^2, \tag{4.2.2}$$

because

$$\begin{aligned}\mathrm{Var}(X)&=E((X-\mu)^2)\\&=E(X^2-2\mu X+\mu^2)\\&=E(X^2)-2\mu E(X)+\mu^2\\&=E(X^2)-[E(X)]^2.\end{aligned}$$

Example 4.2.1

Suppose that $X\sim B(1,p)$ with $p\in(0,1)$, find $\mathrm{Var}(X)$.

Solution. Since $X\sim B(1,p)$, we have $X^2\sim B(1,p)$ too. This follows that $E(X)=E(X^2)=p$. So,

$$\mathrm{Var}(X)=E(X^2)-[E(X)]^2=p-p^2=p(1-p).$$

Example 4.2.2

Suppose that X has a Poisson distribution with λ, that is $X \sim P(\lambda)$, find $\mathrm{Var}(X)$.

Solution.

By Example 4.1.4, $E(X)=\lambda$. Next we compute $E(X^2)$. Since

$$P\{X=k\}=\frac{\lambda^k}{k!}\mathrm{e}^{-\lambda}, \quad k=0,1,2,\cdots,$$

we have

$$\begin{aligned} E(X^2) &= E(X(X-1)+X) \\ &= E(X(X-1))+E(X) \\ &= \sum_{k=0}^{\infty} k(k-1)\frac{\lambda^k}{k!}\mathrm{e}^{-\lambda}+\lambda \\ &= \sum_{k=2}^{\infty} k(k-1)\frac{\lambda^k}{k!}\mathrm{e}^{-\lambda}+\lambda \\ &= \lambda^2\mathrm{e}^{-\lambda}\sum_{k=2}^{\infty}\frac{\lambda^{k-2}}{(k-2)!}+\lambda \\ &= \lambda^2+\lambda. \end{aligned}$$

Hence,

$$\mathrm{Var}(X)=E(X^2)-[E(X)]^2=\lambda^2+\lambda-\lambda^2=\lambda.$$

Example 4.2.3

Suppose that X has a uniform distribution on (a,b) with $a<b$, that is $X \sim U(a,b)$, find $\mathrm{Var}(X)$.

Solution.

We firstly note that the p. d. f. is

$$f(x)=\begin{cases} \frac{1}{b-a}, & x\in(a,b); \\ 0, & \text{otherwise.} \end{cases}$$

and the expectation $E(X)=\frac{a+b}{2}$. Since

$$E(X^2) = \int_{-\infty}^{+\infty} x^2 f(x) \mathrm{d}x$$
$$= \int_a^b x^2 \frac{1}{b-a} \mathrm{d}x$$
$$= \frac{a^2 + ab + b^2}{3},$$

we have

$$\mathrm{Var}(X) = E(X^2) - [E(X)]^2$$
$$= \frac{a^2 + ab + b^2}{3} - \left(\frac{a+b}{2}\right)^2$$
$$= \frac{(b-a)^2}{12}.$$

Example 4. 2. 4

Suppose that X has an exponential distribution with parameter $\lambda > 0$, its p. d. f. is

$$f(x) = \begin{cases} \lambda e^{-\lambda x}, & x > 0, \\ 0, & \text{otherwise}. \end{cases}$$

Find $\mathrm{Var}(X)$.

Solution.

We firstly note that $E(X) = 1/\lambda$. Since

$$E(X^2) = \int_{-\infty}^{+\infty} x^2 f(x) \mathrm{d}x = \int_0^{+\infty} x^2 \lambda e^{-\lambda x} \mathrm{d}x = \frac{2}{\lambda^2}.$$

So,

$$\mathrm{Var}(X) = E(X^2) - [E(X)]^2 = \frac{2}{\lambda^2} - \left(\frac{1}{\lambda}\right)^2 = \frac{1}{\lambda^2}.$$

Example 4. 2. 5

Suppose that $X \sim N(0,1)$. Find $\mathrm{Var}(X)$.

Solution.

Since $X \sim N(0,1)$, its p. d. f. is

$$f(x) = \frac{1}{\sqrt{2\pi}} e^{-\frac{x^2}{2}}, \quad -\infty < x < \infty.$$

Then

$$\begin{aligned}
\mathrm{Var}(X) &= E(X^2) - [E(X)]^2 \\
&= E(X^2) \\
&= \int_{-\infty}^{+\infty} x^2 f(x)\mathrm{d}x \\
&= \frac{1}{\sqrt{2\pi}}\int_{-\infty}^{+\infty} x^2 \mathrm{e}^{-\frac{x^2}{2}}\mathrm{d}x \\
&= \frac{1}{\sqrt{2\pi}}\int_{-\infty}^{+\infty} -x\mathrm{d}\mathrm{e}^{-\frac{x^2}{2}} \\
&= \frac{1}{\sqrt{2\pi}}\left[-x\mathrm{e}^{-\frac{x^2}{2}}\Big|_{-\infty}^{+\infty} + \int_{-\infty}^{+\infty} \mathrm{e}^{-\frac{x^2}{2}}\mathrm{d}x\right] \\
&= 1.
\end{aligned}$$

4.2.2 Properties of Variance

We now present several theorems pertaining to the basic properties of the variance. In these theorems we shall assume that the variances of all the random variables exist.

Theorem 4.2.1

$\mathrm{Var}(X)=0$ if and only if there exists a constant c such that $P\{X=c\}=1$.

Solution.

We first prove the necessity. Suppose first that there exists a constant c such that $P\{X=c\}=1$. Then $E(X)=c$, and $P\{(X-c)^2=0\}=1$. Therefore,

$$\mathrm{Var}(X)=E((X-c)^2)=0.$$

Next we prove the sufficiency. Suppose that $\mathrm{Var}(X)=0$. Then $P\{(X-\mu)^2\geqslant 0\}=1$ but $E((X-\mu)^2)=0$. Therefore, in accordance with Proposition 4.1.5, it can be seen that

$$P\{(X-\mu)^2=0\}=1.$$

Hence, $P\{X=\mu\}=1$. That is, $\mu=c$.

Theorem 4.2.2

For every constants a and b,

$$\mathrm{Var}(aX+b)=a^2\mathrm{Var}(X).$$

Proof.

If $E(X)=\mu$, then $E(aX+b)=a\mu+b$. Therefore,

$$\begin{aligned}\operatorname{Var}(aX+b)&=E((aX+b-a\mu-b)^2)\\&=E((aX-a\mu)^2)\\&=a^2E((X-\mu)^2)\\&=a^2\operatorname{Var}(X).\end{aligned}$$

In Theorem 4.2.2, if $a=0$, then $\operatorname{Var}(b)=0$. That is, the variance for constant is zero.

Example 4.2.6

Suppose that $X\sim N(\mu,\sigma^2)$, then find $\operatorname{Var}(X)$.

Solution.

Assume that $Y\sim N(0,1)$, then $\sigma Y+\mu\sim N(\mu,\sigma^2)$. By Theorem 4.2.2,

$$\operatorname{Var}(X)=\operatorname{Var}(\sigma X+\mu)=\sigma^2.$$

Theorem 4.2.3

If X_1, X_2 are 2 independent random variables such that each variance $\operatorname{Var}(X_i)$ exists, $i=1,2$, then

$$\operatorname{Var}(X_1+X_2)=\operatorname{Var}(X_1)+\operatorname{Var}(X_2). \tag{4.2.3}$$

Proof.

Suppose $E(X_1)=\mu_1$ and $E(X_2)=\mu_2$, then

$$E(X_1+X_2)=\mu_1+\mu_2.$$

Therefore,

$$\begin{aligned}\operatorname{Var}(X_1+X_2)&=E((X_1+X_2-\mu_1-\mu_2)^2)\\&=E((X_1-\mu_1)^2+(X_2-\mu_2)^2+2(X_1-\mu_1)(X_2-\mu_2))\\&=\operatorname{Var}(X_1)+\operatorname{Var}(X_2)+2E((X_1-\mu_1)(X_2-\mu_2)).\end{aligned}$$

Since X_1 and X_2 are independent,

$$\begin{aligned}E((X_1-\mu_1)(X_2-\mu_2))&=E(X_1-\mu_1)E(X_2-\mu_2)\\&=(\mu_1-\mu_1)(\mu_2-\mu_2)\\&=0.\end{aligned}$$

It follows, therefore, that

$$\operatorname{Var}(X_1+X_2)=\operatorname{Var}(X_1)+\operatorname{Var}(X_2).$$

Remark 4.2.2

We note, the formula (4.2.3) can be generalized to the case of n independent random variables: If $X_1, \cdots, X_n$ are n independent random variables such that each expectation $\mathrm{Var}(X_i)$ exists ($i=1,\cdots,n$), then

$$\mathrm{Var}(X_1+\cdots+X_n)=\mathrm{Var}(X_1)+\cdots+\mathrm{Var}(X_n). \qquad (4.2.4)$$

Example 4.2.7

Suppose that $X\in B(n,p)$ with $p\in(0,1)$, compute $\mathrm{Var}(X)$.

Solution.

Suppose that $X_i \sim B(1,p), i=1,2,\cdots,n$, and they are independent, then

$$\mathrm{Var}(X_i)=p(1-p), \quad i=1,2,\cdots,n.$$

So, we can assume that

$$X_1+X_2+\cdots+X_n \sim B(n,p).$$

It follows from (4.16) that

$$\begin{aligned}\mathrm{Var}(X) &= \mathrm{Var}(X_1+X_2+\cdots+X_n)\\ &= \mathrm{Var}(X_1)+\mathrm{Var}(X_2)+\cdots+\mathrm{Var}(X_n)\\ &= np(1-p).\end{aligned}$$

Example 4.2.8 (the variance in Example 4.1.12)

Suppose that a person types n letters and n envelopes, and then places n letters into the n envelopes in a random manner with that one envelop is put one letter only. Let X be the number of letters that are placed in the correct envelopes. Find the variance of X.

Solution.

For $i=1, \cdots, n$, let $X_i=1$ if the ith letter is placed in the correct envelope, and $X_i=0$ otherwise. Then $X=X_1+\cdots+X_n$. However $X_1,\cdots,X_n$ are not independent.

It is said in Example 4.1.12. that $E(X)=1$. Since

$$\begin{aligned}E(X^2) &= E(X_1+\cdots+X_n)^2\\ &= E\Big(\sum_{i=1}^{n} X_i^2 + 2\sum_{1\leqslant i<j\leqslant n}(X_iX_j)\Big)\\ &= \sum_{i=1}^{n} E(X_i^2)+2\sum_{1\leqslant i<j\leqslant n} E(X_iX_j),\end{aligned}$$

where

$$E(X_i^2)=P\{X_i^2=1\}=P\{X_i=1\}=\frac{1}{n}, \quad i=1,2,\cdots,n,$$

$$E(X_iX_j)=P\{X_iX_j=1\}=\frac{1}{n(n-1)}, \quad i,j=1,2,\cdots,n,$$

it follows that

$$E(X^2)=1+n(n-1)\frac{1}{n(n-1)}=2.$$

Hence,

$$\mathrm{Var}(X)=E(X^2)-(EX)^2=1.$$

4.2.3 Chebyshev Inequility

Theorem 4.2.4 (the Chebyshev Inequality)

Let X be a random variable for which $\mathrm{Var}(X)$ exists. Then for every number $t>0$

$$P\{|X-E(X)|\geqslant t\}\leqslant\frac{\mathrm{Var}(X)}{t^2}. \tag{4.2.5}$$

Proof.

Let $Y=[X-E(X)]^2$. Then $P\{Y\geqslant 0\}=1$ and $E(Y)=\mathrm{Var}(X)$. By applying the Markov inequality to Y, we obtain the following result:

$$P\{|X-E(X)|\geqslant t\}=P\{Y\geqslant t^2\}\leqslant\frac{\mathrm{Var}(X)}{t^2}.$$

It is easy to find that formula (4.2.5) has an equivalent form:

$$P\{|X-E(X)|<t\}\geqslant 1-\frac{\mathrm{Var}(X)}{t^2}. \tag{4.2.6}$$

It can be seen from this proof that the Chebyshev inequality is simply a special case of the Markov inequality. Therefore, the comments that were given following the proof of the Markov inequality can be applied as well to the Chebyshev inequality. Because of their generality, these inequalities are very useful. For example, if $\mathrm{Var}(X)=\sigma^2$ and we let $t=3\sigma$, then the Chebyshev inequality yields the result that

$$P(|X-E(X)|\geqslant 3\sigma)\leqslant\frac{1}{9},$$

$$P(|X-E(X)|<3\sigma)\geqslant 1-\frac{1}{9}=\frac{8}{9}.$$

In words, this result says that the probability that any given random variable will differ from its mean value by more than 3 standard deviations does not exceed 1/9, or the probability that the random variable will be in the interval with radius 3 around the mean value is more than 8/9.

Example 4.2.9

Suppose that $\mathrm{Var}(X)=0$, then prove $P\{X=E(X)\}=1$.

We note that the result follows from Theorem 4.2.1 and omit the proof.

4.2.4 Moment

We first give the definition of moment and then prove the Cauchy-Schwarz inequality.

Definition 4.2.2

Suppose X,Y are two random variables, k is any positive integer.

(1) If $E(|X^k|)<\infty$, $E(X^k)$ is called the kth original moment of X.

(2) If $E(X)=\mu$ is finite and $E(|X-\mu|^k)<\infty$, $E((X-\mu)^k)$ is called the kth central moment of X or the kth moment of X about the mean.

(3) If $E(|X^kY^l|)<\infty$, $E(X^kY^l)$ is called the $k+l$ original moment of X and Y.

(4) If $E(X)=\mu, E(Y)=\lambda$ are finite and $E(|X-\mu|^k|Y-\lambda|^l)<\infty$, $E((X-\mu)^k(Y-\lambda)^l)$ is called the $k+l$ central moment of X and Y, or the $k+l$th moment of X and Y about their means.

By Definition 4.2.2, the expectation of X is the first moment of X, the variance of X is the second central moment of X.

Example 4.2.10

If $E(|X|^k)<\infty$ for some positive integer k, then $E(|X|^j)<\infty$ for every positive integer j such that $j<k$.

Proof.

For convenience, we suppose that the distribution of X is continuous and the p. d. f. is f. Then

$$\begin{aligned} E(|X|^j) &= \int_{-\infty}^{+\infty} |x|^j f(x)\mathrm{d}x \\ &= \int_{|x|\leqslant 1} |x|^j f(x)\mathrm{d}x + \int_{|x|>1} |x|^j f(x)\mathrm{d}x \\ &\leqslant \int_{|x|\leqslant 1} 1 \cdot f(x)\mathrm{d}x + \int_{|x|>1} |x|^k f(x)\mathrm{d}x \\ &\leqslant P(|X|\leqslant 1) + E(|X|^k). \end{aligned}$$

By hypothesis, $E(|X|^k)<\infty$, it follows that $E(|X|^j)<\infty$. A similar proof holds for a discrete or a more general type of distribution.

Theorem 4. 2. 5 (Cauchy-Schwarz Inequality)

Suppose that the second moment of X and Y are not zeros and are finite. Then

$$[E(XY)]^2 \leqslant E(X^2)E(Y^2), \tag{4. 2. 7}$$

and formula (4. 2. 7) is equality if and only if there exists a constant a such that $P\{Y=aX\}=1$.

Proof.

Define function $g(t)=E((Y+tX)^2)$. We have

$$g(t)=E(X^2)t^2+2E(XY)t+E(Y^2)\geqslant 0 \text{ for all } t.$$

Since $E(X^2)>0$, it follows from the property of quadratic equation that

$$[2E(XY)]^2-4E(X^2)E(Y^2)\leqslant 0,$$

which is equivalent to

$$[E(XY)]^2 \leqslant E(X^2)E(Y^2).$$

Suppose that

$$[E(XY)]^2 = E(X^2)E(Y^2),$$

that is,

$$[2E(XY)]^2-4E[X^2]E(Y^2)=0,$$

which means that there is a unique t_0 such that

$$g(t_0)=E((Y+t_0X)^2)=0.$$

However

$$0=E((Y+t_0X)^2)=\mathrm{Var}(Y+t_0X)+[E(Y+t_0X)]^2,$$

which gives

$$\mathrm{Var}(Y+t_0X)=0, \quad E(Y+t_0X)]=0.$$

By Theorem 4.2.1, this is equivalent to $P\{Y=-t_0X\}=1$. Hence, the result holds by letting $a=-t_0$.

4.3 Covariance and Correlation

When we study the joint distribution of two random variables, it is useful to summarize how much the two random variables depend on each other. The covariance and correlation are two tools that measure the dependence, but different from the conditional d. f. s they only capture a particular type of dependence, namely linear dependence.

Let X and Y be random variables having a specified joint distribution, and let $E(X)=\mu_X$, $E(Y)=\mu_Y$, $\mathrm{Var}(X)=\sigma_X^2$, and $\mathrm{Var}(Y)=\sigma_Y^2$. It says in Theorem 4.2.3 that if X and Y are independent, then

$$E((X-E(X))(Y-E(Y)))=0,$$

and it follows that $\mathrm{Var}(X+Y)=\mathrm{Var}(X)+\mathrm{Var}(Y)$. However, if X and Y are not independent, then

$$\mathrm{Var}(X+Y)=\mathrm{Var}(X)+\mathrm{Var}(Y)+2E((X-\mu_X)(Y-\mu_Y)).$$

In summary, we can claim that the term $E((X-\mu_X)(Y-\mu_Y))$ embodies some relation between X and Y when they are not independent.

Definition 4.3.1

The covariance of X and Y, denoted by $\mathrm{Cov}(X,Y)$, is defined as

$$\mathrm{Cov}(X,Y)=E((X-\mu_X)(Y-\mu_Y)). \tag{4.3.1}$$

If $0<\sigma_X<+\infty$ and $0<\sigma_Y<+\infty$, then the correlation of X and Y, denoted by $\rho(X,Y)$, is defined as

$$\rho(X,Y)=\frac{\mathrm{Cov}(X,Y)}{\sqrt{\mathrm{Var}(X)}\sqrt{\mathrm{Var}(Y)}}=\frac{\mathrm{Cov}(X,Y)}{\sigma_X\sigma_Y}, \tag{4.3.2}$$

Where $\sigma_X = \sqrt{Var(X)}$, and $\sigma_Y = \sqrt{Var(Y)}$ are standard deviations of X and Y respectively.

From Definition 4. 3. 1, it says that $Cov(X,Y)$ is a function of X and Y. So one can compute it by (4. 1. 6) for discrete random variables or (4. 1. 7) for continuous random variables. In words, the covariance only depends on the joint d. f. of X and Y.

However we often use the following formula. If $\sigma_X < +\infty$ and $\sigma_Y < +\infty$, then

$$Cov(X,Y) = E(XY) - E(X)E(Y), \tag{4.3.3}$$

which holds because

$$\begin{aligned} Cov(X,Y) &= E(XY - \mu_X Y - \mu_Y X + \mu_X \mu_Y) \\ &= E(XY) - \mu_X E(Y) - \mu_Y E(X) + \mu_X \mu_Y \\ &= E(XY) - E(X)E(Y). \end{aligned}$$

Example 4. 3. 1

Suppose that continuous random variable X has a p. d. f.

$$f(x) = \begin{cases} 2x, & x \in (0,1); \\ 0, & \text{otherwise}. \end{cases}$$

And $Y = X^2$. Find $Cov(X,Y)$ and $\rho(X,Y)$.

Solution.

Because $E(X) = 2/3$, $E(Y) = E(X^2) = 1/2$, and $E(XY) = E(X^3) = 2/5$, by formula (4. 3. 3) we have,

$$Cov(X,Y) = E(XY) - E(X)E(Y) = \frac{2}{5} - \frac{1}{3} = \frac{1}{15}.$$

For the correlation, since

$$Var(X) = E(X^2) - [E(X)]^2 = \frac{1}{2} - \left(\frac{2}{3}\right)^2 = \frac{1}{18},$$

$$Var(Y) = E(Y^2) - [E(Y)]^2 = E(X^4) - [E(X^2)] = \frac{1}{3} - \left(\frac{1}{2}\right)^2 = \frac{1}{12},$$

we have

$$\rho(X,Y) = \frac{Cov(X,Y)}{\sqrt{Var(X)}\sqrt{Var(Y)}} = \frac{2\sqrt{6}}{5}.$$

Next we show the properties of the covariance and correlation.

Proposition 4.3.1

The covariance satisfies the following:

(1) $\mathrm{Cov}(X,X)=\mathrm{Var}(X)$;

(2) $\mathrm{Cov}(aX,bY)=ab\mathrm{Cov}(Y,X)$ for any constants a and b;

(3) $\mathrm{Cov}(X_1+X_2,Y)=\mathrm{Cov}(X_1,Y)+\mathrm{Cov}(X_2,Y)$.

The proofs are omitted.

Example 4.3.2

Suppose $X_1,\cdots,X_n$ are random variables such that $\mathrm{Var}(X_i)<+\infty$ for $i=1,\cdots,n$. Then

$$\mathrm{Var}\left(\sum_{i=1}^{n}X_i\right)=\sum_{i=1}^{n}\mathrm{Var}(X_i)+2\sum\sum_{i<j}\mathrm{Cov}(X_i,X_j). \quad (4.3.4)$$

Solution.

For every random variable X, $\mathrm{Cov}(X,X)=\mathrm{Var}(X)$. Therefore, by Proposition 4.3.1 we can obtain the following relation:

$$\mathrm{Var}\left(\sum_{i=1}^{n}X_i\right)=\mathrm{Cov}\left(\sum_{i=1}^{n}X_i,\sum_{j=1}^{n}X_j\right)=\sum_{i=1}^{n}\sum_{j=1}^{n}\mathrm{Cov}(X_i,X_j).$$

We shall separate the final sum in this relation into two sums: the sum of those terms for which $i=j$ and the sum of those terms for which $i\neq j$. Then, if we use the fact that $\mathrm{Cov}(X_i,X_j)=\mathrm{Cov}(X_j,X_i)$, we obtain the relation

$$\begin{aligned}\mathrm{Var}\left(\sum_{i=1}^{n}X_i\right)&=\sum_{i=1}^{n}\mathrm{Var}(X_i)+\sum\sum_{i\neq j}\mathrm{Cov}(X_i,X_j)\\&=\sum_{i=1}^{n}\mathrm{Var}(X_i)+2\sum\sum_{i<j}\mathrm{Cov}(X_i,X_j).\end{aligned}$$

As a special case of Example 4.3.2, we often use the following, for random variables X,Y and constants a,b,c,

$$\mathrm{Var}(X\pm Y)=\mathrm{Var}(X)+\mathrm{Var}(Y)\pm 2\mathrm{Cov}(X,Y),$$

$$\mathrm{Var}(aX+bY+c)=a^2\mathrm{Var}(X)+b^2\mathrm{Var}(Y)+2ab\mathrm{Cov}(X,Y).$$

Proposition 4.3.2

Suppose that $0<\sigma_X<+\infty$ and $0<\sigma_Y<+\infty$, the correlation $\rho(X,Y)$ satisfies the following:

(1) $|\rho(X,Y)|\leqslant 1$;

(2) $|\rho(X,Y)|=1$ if and only if there exist constant a and b such that $P\{Y=aX+b\}=1$.

Proof.

(1) Let $U=X-E(X)$ and $V=Y-E(Y)$, by formula (4.2.7) in Theorem 4.2.5, we have

$$|\rho(X,Y)|=\left|\frac{\mathrm{Cov}(X,Y)}{\sqrt{\mathrm{Var}(X)}\sqrt{\mathrm{Var}(Y)}}\right|=\left|\frac{E(UV)}{\sqrt{E(U^2)}\sqrt{E(V^2)}}\right|\leqslant 1.$$

(2) From above, $|\rho(X,Y)|=1$ if and only if

$$E((UV)^2)=E(U^2)E(V^2),$$

which, by Theorem 4.2.5, is equivalent to that there exists a constant t_0 such that

$$P\{V=t_0U\}=1,$$

that is,

$$P\{Y=t_0X-t_0E(X)+E(Y)\}=1.$$

So, the result holds with $a=t_0$ and $b=-t_0E(X)+E(Y)$.

The next result is a corollary to Proposition 4.3.2.

Corollary 4.3.1

Suppose that X is a random variable such that $0<\sigma_X^2<\infty$. Let $Y=aX+b$ for some constants a and b, where $a\neq 0$. Then

$$\rho(X,Y)=\begin{cases}1, & a>0;\\ -1, & a<0.\end{cases}$$

Proof.

If $Y=aX+b$, then $\mu_Y=a\mu_X+b$ and $Y-\mu_Y=a(X-\mu_X)$. Therefore,

$$\mathrm{Cov}(X,Y)=aE((X-\mu_X)^2)=a\sigma_X^2.$$

Since $\sigma_Y=|a|\sigma_X$, the theorem follows from formula (4.3.2).

There is a converse to Corollary 4.3.1. That is, $|\rho(X,Y)|=1$ implies that X and Y are linearly related. In general, the value of $\rho(X,Y)$ provides a measure of the extent to which two random variables X and Y are linearly related. If the joint distribution of X and Y is relatively concentrated around a straight line in the xy-plane that has a positive slope, then $\rho(X,Y)$ will typically be close to 1. If the joint distribution is relatively concentrated around a straight line that has a negative slope, then $\rho(X,Y)$ will typically be close to -1. We shall not discuss these concepts further here, but we shall consider them again when the bivariate normal distribution is introduced.

A large value of $|\rho(X,Y)|$ means that X and Y are close to being linearly related, and hence are closely related. But a small value of $|\rho(X,Y)|$ does not mean that X and Y are not close to being related.

Definition 4.3.2

If $\rho(X,Y)=0$, then X and Y are said to be uncorrelated.

In fact, for the uncorrelated we have the following equivalence:

(i) $\rho(X,Y)=0$;

(ii) $\mathrm{Cov}(X,Y)=0$;

(iii) $E(X,Y)-E(X)E(Y)=0$;

(iv) $\mathrm{Var}(X\pm Y)=\mathrm{Var}(X)+\mathrm{Var}(Y)$.

In view of Proposition 4.1.3, the independence of X and Y implies the uncorrelated if the correlation exists, that is, if X and Y are independent with standard deviations $0<\sigma_X<+\infty$ and $0<\sigma_Y<+\infty$, then

$$\mathrm{Cov}(X,Y)=E(XY)-E(X)E(Y)=0,$$

and $\rho(X,Y)=0$. So, if X and Y are independent then (ii) (iii) and (iv) above hold too. However, the uncorrelated does not generally implies the independence except for the bivariate normal distribution.

Example 4.3.3

Suppose that (X,Y) has a bivariate normal distribution with p. d. f.

$$f(x,y)=\frac{1}{2\pi\sigma_1\sigma_2\sqrt{1-\rho^2}}\cdot$$

$$\exp\left\{-\frac{1}{2(1-\rho^2)}\left[\left(\frac{x-\mu_1}{\sigma_1}\right)^2-2\rho\left(\frac{x-\mu_1}{\sigma_1}\right)\left(\frac{y-\mu_2}{\sigma_2}\right)+\left(\frac{y-\mu_2}{\sigma_2}\right)\right]^2\right\}.$$

Find the correlation $\rho(X,Y)$ and prove that the uncorrelated, $\rho(X,Y)=0$, is equivalent to the independence of X and Y.

Solution.

(i) By Example 3.2.4, we know the marginal $X \sim N(\mu_1, \sigma_1^2)$, $Y \sim N(\mu_2, \sigma_2^2)$, the marginal p. d. f. s are

$$f_X(x) = \frac{1}{\sqrt{2\pi}\sigma_1} e^{-\frac{(x-\mu_1)^2}{2\sigma_1^2}}, \quad -\infty < x < +\infty,$$

$$f_Y(y) = \frac{1}{\sqrt{2\pi}\sigma_2} e^{-\frac{(y-\mu_2)^2}{2\sigma_2^2}}, \quad -\infty < y < +\infty,$$

and $E(X) = \mu_1$, $E(Y) = \mu_2$, $\mathrm{Var}(X) = \sigma_1^2$ and $\mathrm{Var}(Y) = \sigma_2^2$. Next we compute the covariance,

$$\begin{aligned}
\mathrm{Cov}(X,Y) &= \int_{-\infty}^{+\infty}\int_{-\infty}^{+\infty} (x-\mu_1)(y-\mu_2) f(x,y) \mathrm{d}x\mathrm{d}y \\
&= \frac{1}{2\pi\sigma_1\sigma_2\sqrt{1-\rho^2}} \int_{-\infty}^{+\infty}\int_{-\infty}^{+\infty} (x-\mu_1)(y-\mu_2) \\
&\cdot \exp\left[\frac{-1}{2(1-\rho^2)}\left(\frac{y-\mu_2}{\sigma_2} - \rho\frac{x-\mu_1}{\sigma_1}\right)^2 - \frac{(x-\mu_1)^2}{2\sigma_1^2}\right] \mathrm{d}y\mathrm{d}x.
\end{aligned}$$

Let

$$t = \frac{1}{\sqrt{(1-\rho^2)}}\left(\frac{y-\mu_2}{\sigma_2} - \rho\frac{x-\mu_1}{\sigma_1}\right), \quad u = \frac{x-\mu_1}{\sigma_1},$$

Then

$$\begin{aligned}
\mathrm{Cov}(X,Y) &= \frac{1}{2\pi}\int_{-\infty}^{+\infty}\int_{-\infty}^{+\infty} (\sigma_1\sigma_2\sqrt{1-\rho^2}\, tu + \rho\sigma_1\sigma_2 u^2)\, e^{-(u^2+t^2)/2} \mathrm{d}t\mathrm{d}u \\
&= \frac{\rho\sigma_1\sigma_2}{2\pi}\int_{-\infty}^{+\infty} u^2 e^{-u^2/2} \mathrm{d}u \cdot \int_{-\infty}^{+\infty} e^{-t^2/2} \mathrm{d}t \\
&\quad + \frac{\sigma_1\sigma_2\sqrt{1-\rho^2}}{2\pi}\int_{-\infty}^{+\infty} u e^{-u^2/2} \mathrm{d}u \cdot \int_{-\infty}^{+\infty} t e^{-t^2/2} \mathrm{d}t \\
&= \frac{\rho\sigma_1\sigma_2}{2\pi}\sqrt{2\pi}\cdot\sqrt{2\pi} = \rho\sigma_1\sigma_2.
\end{aligned}$$

So, $\rho(X,Y) = \rho$ by definition.

(ii) By Example 3.4.3, if $\rho(X,Y) = 0$, that is, X and Y are uncorrelated, then they are independent. Hence, the uncorrelated, $\rho(X,Y) = 0$, is equivalent to their independence.

The next example gives two random variables which are not only dependent but also uncorrelated.

Example 4.3.4

Suppose that Z has a uniform distribution on $(-\pi, \pi)$, that is, $Z \sim U(-\pi, \pi)$. Let $X = \sin Z$ and $Y = \cos Z$. Verify X and Y are both dependent and uncorrelated.

Solution.

Since the p. d. f. of Z is

$$f_Z(z) = \begin{cases} \dfrac{1}{2\pi}, & x \in (-\pi, \pi); \\ 0, & \text{otherwise.} \end{cases}$$

we have

$$E(X) = \frac{1}{2\pi}\int_{-\pi}^{\pi} \sin z \mathrm{d}z = 0,$$

$$E(Y) = \frac{1}{2\pi}\int_{-\pi}^{\pi} \cos z \mathrm{d}z = 0,$$

$$E(X^2) = \frac{1}{2\pi}\int_{-\pi}^{\pi} (\sin z)^2 \mathrm{d}z = \frac{1}{2},$$

$$E(Y^2) = \frac{1}{2\pi}\int_{-\pi}^{\pi} (\cos z)^2 \mathrm{d}z = \frac{1}{2},$$

$$E(XY) = \frac{1}{2\pi}\int_{-\pi}^{\pi} \sin z \cos z \mathrm{d}z = 0.$$

This follows $\mathrm{Cov}(X, Y) = 0$, and $\rho(X, Y) = 0$. That is, X and Y are uncorrelated.

Notice that $X^2 + Y^2 = (\sin Z)^2 + (\cos Z)^2 = 1$, we claim that X and Y are not independent because

$$P\{0 < X < \frac{1}{2}\} P\{0 < Y < \frac{1}{2}\} \neq 0 = P\{0 < X < \frac{1}{2}, 0 < Y < \frac{1}{2}\}.$$

4.4 Covariance Matrix

When we study the joint distribution of two and more random variables, it is easier to use matrix form.

Next we will present the mean vector and covariance matrix. Multi-normal distribution will be given as an example.

Let $X=(X_1, X_2, \cdots, X_n)'$ is a n-dimensional random variable with mean $\mu_i = E(X_i)$ and covariance $C_{ij} = \text{Cov}(X_i, X_j)$, $i, j = 1, 2, \cdots, n$. Let mean vector

$$\mu \equiv E(X) = (\mu_1, \mu_2, \cdots, \mu_n)'$$

and covariance matrix

$$\mathbf{C} = \begin{pmatrix} C_{11} & C_{12} & \cdots & C_{1n} \\ C_{21} & C_{22} & \cdots & C_{2n} \\ \vdots & \vdots & & \vdots \\ C_{n1} & C_{n2} & \cdots & C_{nn} \end{pmatrix},$$

where $C_{ii} = \text{Var}(X_i)$ for $i=1,2,\cdots,n$.

The covariance matrix $\mathbf{C}$ has the following properties:

(i) $\mathbf{C}$ is symmetrical matrix, that is, $\mathbf{C}=\mathbf{C}'$;

(ii) $\mathbf{C}$ is nonnegative definite, that is, for any $t=(t_1, t_2, \cdots, t_n)'$ we have

$$t'\mathbf{C}t \geqslant 0.$$

It is easy to see that (i) holds. For (ii),

$$\begin{aligned} t'\mathbf{C}t &= \sum_{i=1}^{n}\sum_{j=1}^{n} C_{ij} t_i t_j \\ &= \sum_{i=1}^{n}\sum_{j=1}^{n} t_i E(X_i - \mu_i)(X_j - \mu_j) t_j \\ &= E\Big(\sum_{j=1}^{n} t_i (X_i - \mu_i)\Big)^2 \\ &\geqslant 0. \end{aligned}$$

Next we use the mean vector and covariance matrix to formula the 2-dimensional normal p. d. f.

Suppose that $X=(X_1, X_2)' \sim N(\mu_1, \mu_2, \sigma_1^2, \sigma_2^2, \rho)$ and $|\rho| \neq 1$. We have

$$\mu = E(X) = (\mu_1, \mu_2)',$$

$$\mathbf{C} = \begin{pmatrix} \sigma_1^2 & \rho\sigma_1\sigma_2 \\ \rho\sigma_1\sigma_2 & \sigma_2^2 \end{pmatrix}$$

with the determinant

$$\det(\boldsymbol{C})=\sigma_1^2\sigma_2^2(1-\rho^2).$$

Since $|\rho|\neq 1$, the matrix $\boldsymbol{C}$ is nonsingular and its inverse is

$$\boldsymbol{C}^{-1}=\frac{1}{\sigma_1^2\sigma_2^2(1-\rho^2)}\begin{pmatrix}\sigma_2^2 & -\rho\sigma_1\sigma_2\\ -\rho\sigma_1\sigma_2 & \sigma_1^2\end{pmatrix}.$$

Letting $x=(x_1,x_2)'$ we have

$$\begin{aligned}&(x-\mu)'\boldsymbol{C}^{-1}(x-\mu)\\ &=\frac{1}{1-\rho^2}(x_1-\mu_1,x_2-\mu_2)\begin{pmatrix}\dfrac{1}{\sigma_1^2} & -\dfrac{\rho}{\sigma_1\sigma_2}\\ -\dfrac{\rho}{\sigma_1\sigma_2} & \dfrac{1}{\sigma_2^2}\end{pmatrix}\begin{pmatrix}x_1-\mu_1\\ x_2-\mu_2\end{pmatrix}\\ &=\frac{1}{1-\rho^2}\left[\frac{(x_1-\mu_1)^2}{\sigma_1^2}-2\rho\frac{(x_1-\mu_1)(x_2-\mu_2)}{\sigma_1\sigma_2}+\frac{(x_2-\mu_2)^2}{\sigma_2^2}\right].\end{aligned}$$

So, associated with Example 4. 3. 3, the joint p. d. f. $f(x_1,x_2)$ has the following matrix form

$$f(x_1,x_2)=\frac{1}{(2\pi)^{2/2}|\det(\boldsymbol{C})|^{1/2}}\mathrm{e}^{-\frac{1}{2}(x-\mu)'\boldsymbol{C}^{-1}(x-\mu)}.$$

In general, the p. d. f. of the n-dimensional normal random variable is defined as

$$f(x_1,x_2,\cdots,x_n)=\frac{1}{(2\pi)^{n/2}|\det(\boldsymbol{C})|^{1/2}}\mathrm{e}^{-\frac{1}{2}(x-\mu)'\boldsymbol{C}^{-1}(x-\mu)},$$

where the mean vector

$$\mu=(\mu_1,\mu_2,\cdots,\mu_n)'$$

and the covariance matrix

$$\boldsymbol{C}=\begin{pmatrix}\sigma_1^2 & \rho\sigma_1\sigma_2 & \cdots & \rho\sigma_1\sigma_n\\ \rho\sigma_1\sigma_2 & \sigma_2^2 & \cdots & \rho\sigma_2\sigma_n\\ \vdots & \vdots & & \vdots\\ \rho\sigma_1\sigma_n & \rho\sigma_2\sigma_n & \cdots & \sigma_n^2\end{pmatrix}.$$

If $X=(X_1,X_2,\cdots,X_n)'$ has distribution with p. d. f. $f(x_1,x_2,\cdots,x_n)$, we denote $X\sim N(\mu,\boldsymbol{C})$.

Exercises

4. 1 Suppose that random variable X has the following p. f.

Exercise Table 4.1.1

X	-1	0	1
P	$\frac{1}{2}$	$\frac{1}{4}$	$\frac{1}{4}$

Find $E(X)$, $E(X^2)$ and $E(X-1)^2$.

4.2 A fair die is rolled 4 times. Calculate the expected sum of the 4 rolls.

4.3 If X and Y are independent uniform random variables $U(0,1)$, compute $E(|X-Y|)$.

4.4 If X and Y are i. i. d. random variables with mean μ and variance σ^2, find $E((X-Y)^2)$.

4.5 Let Z be a standard normal random variable, and for any fixed x, Let

$$X=\begin{cases} Z, & Z>x; \\ 0, & \text{otherwise.} \end{cases}$$

Show that $E(X)=\frac{1}{\sqrt{2\pi}}e^{-x^2/2}$.

4.6 Suppose that X is a discrete random variable and takes only the values $0,1,2,\cdots$. Prove that

$$E(X)=\sum_{n=1}^{\infty}P\{X\geqslant n\}=\sum_{n=0}^{\infty}P\{X>n\}.$$

4.7 Suppose that random variable X has a continuous distribution with d. f. F and p. d. f. f. (1) If $E(X)$ exists, prove

$$\lim_{t\to\infty}x[1-F(x)]=0.$$

(2) If $E(X)$ exists and also $P\{X\geqslant 0\}=1$, prove that

$$E(X)=\int_0^{\infty}[1-F(x)]dx.$$

4.8 Suppose that X is a continuous random variable with the following p. d. f.

$$f(x)=\begin{cases} 2xe^{-x^2}, & x>0; \\ 0, & \text{otherwise.} \end{cases}$$

Let $Y=X^2$. Compute $E(Y)$ in the following two ways: (1) find the p. d. f. of Y and then compute $E(Y)$; (2) By the formula (4.1.5).

4.9 Suppose that X is a exponential random variable with the following p. d. f.

$$f(x)=\begin{cases} e^{-x}, & x>0; \\ 0, & \text{otherwise.} \end{cases}$$

Let $U=2X$, $V=e^{-X}$ and $W=1-e^{-X}$. Compute $E(U)$, $E(V)$ and $E(W)$.

4.10 A fair die is successively rolled. Let X and Y denote the number of rolls Continuously to obtain a 6 and a 5, respectively. Find $E(X)$, $E(X|Y=1)$ and $E(X|Y=5)$.

4.11 Suppose that X and Y have the following joint p. d. f.

$$f(x,y)=\begin{cases} \dfrac{1}{y}e^{-y}, & 0<x<y, \quad 0<y<\infty; \\ 0, & \text{otherwise,} \end{cases}$$

Find $E(X^3|Y=y)$. If the joint p. d. f. is

$$f(x,y)=\begin{cases} \dfrac{1}{y}e^{-y}e^{-x/y}, & x>0, y>0; \\ 0, & \text{otherwise.} \end{cases}$$

What is $E(X^2|Y=y)$.

4.12 Consider 10 trials with each having the same probability of success p, that is $p=P\{\text{success}\}$. Let X denote the total number of successes in these trials. Calculate $E(X)$ and $\mathrm{Var}(X)$.

4.13 Suppose that X has geometric distribution with the p. f. is

$$P\{X=k\}=(1-p)^{k-1}p, \quad k=1,2,\cdots.$$

Compute $E(X)$ and $\mathrm{Var}(X)$.

4.14 Suppose that random variable X takes values in (a,b), prove the inequalities

$$a\leqslant E(X)\leqslant b, \quad \mathrm{Var}(X)\leqslant\left(\frac{b-a}{2}\right)^2.$$

4.15 Suppose that X and Y have the following joint p. f.

Exercise Table 4.15.1

X \ Y	−1	0	1
1	0.2	0.1	0.1
2	0.1	0	0.1
3	0	0.3	0.1

Find (1) $E(X)$, $E(Y)$, $\mathrm{Var}(X)$, $\mathrm{Var}(Y)$; (2) $E\left(\frac{Y}{X}\right)$; (3) $E(X-Y)^2$.

4.16 Suppose that a fair dice is tossed one time. Let X denote the number of the outcome one. Find $E(X)$ and $\mathrm{Var}(X)$. If the dice is tossed ten times, and let Y be the number of the one obtained, what about $E(Y)$ and $\mathrm{Var}(Y)$.

4.17 Suppose that a certain task is repeatedly and independently performed N times for some positive integer N. Suppose that the probability of success of each given trial is $p\in(0,1)$. Let X be the number of successes in the N trials. Find $E(X)$ and $\mathrm{Var}(X)$.

4.18 Suppose that a fair coin is tossed repeatedly until a head is obtained for the rth time, where r is a positive integer, and X is the needed tossing times. (1) What is the p. f. of X? (2) Find $E(X)$ and $\mathrm{Var}(X)$.

4.19 Let X be the number of 1 and Y the number of 6 that occur in n rolls of a fair die. Find ρ_{XY}.

4.20 Suppose that X and Y have the following joint p. d. f.

$$f(x,y)=\begin{cases}\dfrac{1}{\pi}, & x^2+y^2\leqslant 1;\\ 0, & \text{otherwise.}\end{cases}$$

Computer $\mathrm{Cov}(X,Y)$.

4.21 Let $X_1, X_2, \cdots$ be independent with common mean μ and common variance σ^2, and let $Y_n=X_n+X_{n+1}$. For $j\geqslant 0$, find $\rho(Y_n, Y_{n+j})$.

4.22 If X_1, X_2, X_3, X_4 are pairwise uncorrelated random variables each having mean 0 and variance 1, compute the correlations of (1) X_1+X_2 and

X_2+X_3; (2) X_1+X_2 and X_3+X_4.

4.23 Suppose that X has a normal distribution $N(0,\sigma^2)$, and let $Y=|X|$. Prove that X and Y are not only uncorrelated but also not independent.

4.24 Suppose that X and Y have the following joint p. f.

Exercise Table 4.24.1

X \ Y	−1	0	1
−1	$\frac{1}{8}$	$\frac{1}{8}$	$\frac{1}{8}$
0	$\frac{1}{8}$	0	$\frac{1}{8}$
1	$\frac{1}{8}$	$\frac{1}{8}$	$\frac{1}{8}$

Discuss the independence and the correlation of X and Y.

4.25 Suppose that X and Y have a uniform distribution on the region $D=\{(x,y):0\leqslant x\leqslant 1,0\leqslant y\leqslant x\}$, that is, their p. d. f. is

$$f(x,y)=\begin{cases}2, & (x,y)\in D;\\ 0, & \text{otherwise.}\end{cases}$$

Find $\rho(X,Y)$.

4.26 Suppose that X and Y are i. i. d. normal random variables with distribution $N(\mu,\sigma^2)$. Let $U=aX+bY$ and $V=aX-bY$ for constants a and b not both zeros. Find $\rho(X,Y)$.

4.27 Suppose that X and Y have the following joint p. d. f.

$$f(x,y)=\begin{cases}\frac{1}{8}(x+y), & 0<x<2,0<y<2;\\ 0, & \text{otherwise.}\end{cases}$$

Find $\rho(X,Y)$.

4.28 Suppose that X and Y are random variables such that $\mathrm{Var}(X)=9$, $\mathrm{Var}(Y)=4$, $\rho(X,Y)=-1/4$. Determine (1) $\mathrm{Var}(X+Y)$; (2) $\mathrm{Var}(X-2Y+2)$.

4.29 Suppose that X,Y,Z are three random variables such that $E(X)=E(Y)=1,E(Z)=-1,\mathrm{Var}(X)=\mathrm{Var}(Y)=\mathrm{Var}(Z)=1$, $\rho(X,Y)=0,\rho(Y,Z)=$

$1/2, \rho(X,Z)=-1/2$. Let $W=X+Y+Z$. Find $\mathrm{Var}(W)$.

4.30 Suppose that X and Y have the following joint p. d. f.

$$f(x,y)=\begin{cases}\frac{1}{3}(x+y), & 0<x<1,\ 0<y<2;\\ 0, & \text{otherwise.}\end{cases}$$

Find the covariance matrix of X and Y.

4.31 Suppose that the covariance matrix of X, Y and Z is

$$\boldsymbol{C}_{XYZ}=\begin{pmatrix}4 & 2 & -1\\ 2 & 2 & 3\\ -1 & 3 & 6\end{pmatrix}.$$

Let

$$\begin{cases}U=2X+Y+3Z;\\ V=X-2Y+4Z;\\ W=3X+4Y+Z.\end{cases}$$

Determine the covariance matrix of U, V and W, denoted by $\boldsymbol{C}_{UVW}$.

4.32 Suppose that $(X,Y)\sim N(\mu_1, \mu_2, \sigma_1^2, \sigma_2^2, \rho)$. Find b such that $X-bY$ and $X+bY$ are independent.

4.33 Suppose that $X\sim N(0,1)$, and let I, independent of X, be such that $P\{I=1\}=P\{I=0\}=1/2$. Now define Y by

$$Y=\begin{cases}X, & I=1;\\ -X, & I=0.\end{cases}$$

In words, Y is equally likely to equal either X and $-X$. (1) Are X and Y independent? (2) Are I and Y independent? (3) Prove that Y is normal with mean 0 and variance 1. (4) Show that $\mathrm{Cov}(X,Y)=0$.

Chapter 5 Limit Theorem

In general, the most important theoretical results in probability theory are thought as probable limit theorems. Of those, the most important are the following two: the heading laws of large numbers (LLN) and the central limit theorems (CLT). Usually, theorems are considered to be LLN if they are concerned with stating conditions under which the average of a sequence of random variables converges (in some sense) to the excepted average. On the other hand, CLT is concerned with determining conditions under which the sum of large number of random variables has a probability distribution that is approximately normal distributed.

5.1 Law of Large Numbers

We introduce two LLNs named weak law of large numbers (WLLN) and strong law of large numbers (SLLN).

Consider a sequence of experiments tossing a fair coin and define, for $i=1,2,\cdots$,

$$X_i=\begin{cases}1, & \text{if the outcome of the } i\text{th tossing is head;}\\ 0, & \text{otherwise.}\end{cases}$$

For large n, we define the frequency of "head" in the first n tossing as

$$n_h = \frac{1}{n}\sum_{i=1}^{n} X_i.$$

Intuitively the frequency n_h tends to $1/2$ as n goes to infinity, however in what way does the frequency go to $1/2$ mathematically? We show two convergent ways below.

Definition 5.1.1

Suppose X and X_1, X_2, $\cdots$ are random variables on probability space $(S,\mathscr{F},P)$.

(1) If, for any positive number $\varepsilon>0$,

$$\lim_{n\to+\infty} P\{|X_n-X|\geqslant\varepsilon\}=0, \tag{5.1.1}$$

then it says that X_n converges to X in probability, denoted by $X_n\to X(P)$.

(2) If

$$P\{\lim_{n\to+\infty} X_n=X\}=1, \tag{5.1.2}$$

then it says that X_n converges to X with probability one (w. p. 1) or almost surely (a. s.), denoted by $X_n\to X$ w. p. 1 or a. s.

It can be seen that formula (5.1.1) is equivalent to

$$\lim_{n\to+\infty} P\{|X_n-X|<\varepsilon\}=1, \tag{5.1.3}$$

because

$$P\{|X_n-X|<\varepsilon\}=1-P\{|X_n-X|\geqslant\varepsilon\}.$$

Here we only consider the case with X being a constant. By Definition 5.1.1, we will show next how the frequency n_h goes to $1/2$ in probability and almost surely.

Theorem 5.1.1 (WLLN)

Let $X_1, X_2, \cdots$ be a sequence of i. i. d. random variables, each has finite mean $E(X_i)=\mu$. Then, for any $\varepsilon>0$,

$$P\left\{\left|\frac{X_1+X_2+\cdots+X_n}{n}-\mu\right|\geqslant\varepsilon\right\}\to 0 \text{ as } n\to\infty. \tag{5.1.4}$$

Proof.

We shall prove the result only under the additional assumption that the random variables having a finite variance σ^2. Now notice

$$E\left(\frac{X_1+X_2+\cdots+X_n}{n}\right)=\mu$$

and

$$\operatorname{Var}\left(\frac{X_1+X_2+\cdots+X_n}{n}\right)=\frac{\sum_{i=1}^{n}\operatorname{Var}(X_i)}{n^2}=\frac{\sigma^2}{n}.$$

This, with Chebyshev's inequality in Chapter 4, follows

$$P\left\{\left|\frac{X_1+X_2+\cdots+X_n}{n}-\mu\right|\geqslant\varepsilon\right\}\leqslant\frac{\sigma^2}{n\varepsilon^2}\to 0$$

as $n\to\infty$ and the result is proved.

By Theorem 5.1.1, $n_h\to 1/2\ (P)$.

The WLLN was originally proved by James Bernoulli for the special case where X_i are 0,1 (that is, Bernoulli) random variables. His statement and proof of this theorem were presented in his book Ars Conjectandi published in 1713, 8 years after his death, by his nephew Nicholas Bernoulli. The general form of the WLLN presented in Theorem 5.1.1 was proved by the Russian mathematician Khintchine.

As a corollary to Theorem 5.1.1, Bernoulli WLLN is the following.

Corollary 5.1.1 (Bernoulli WLLN)

Suppose that N_n has a binomial distribution with parameter n and p, that is, $N_n\sim B(n,p)$. For any $\varepsilon>0$, then

$$\lim_{n\to\infty}P\left\{\left|\frac{N_n}{n}-p\right|\geqslant\varepsilon\right\}=0.$$

Proof.

Suppose the event satisfies $P(A)=p$. Define i.i.d. random variables

$$\xi_i=\begin{cases}1, & \text{if the event A occurs in the ith experiment;}\\ 0, & \text{otherwise.}\end{cases}$$

So, we can assume that $N_n=\xi_1+\xi_2+\cdots+\xi_n$. Notice that

$$E(\xi_i)=p,\quad \operatorname{Var}(\xi_i)=p(1-p)\leqslant\frac{1}{4},$$

and

$$\frac{N_n}{n} - p = \frac{1}{n}\sum_{k=1}^{n}\xi_k - \frac{1}{n}\sum_{k=1}^{n}E(\xi_k).$$

By Chebyshev's inequality,

$$P\left\{\left|\frac{N_n}{n} - p\right| \geqslant \varepsilon\right\} \leqslant \frac{1}{\varepsilon^2}\mathrm{Var}\left(\frac{N_n}{n}\right) = \frac{1}{n\varepsilon^2}\mathrm{Var}(\xi_i) \leqslant \frac{1}{4n\varepsilon^2} \to 0$$

as $n \to +\infty$.

In fact, In the proof of Theorem 5.1.1, it can be seen that formula (5.1.4) holds if

$$\frac{1}{n^2}\mathrm{Var}\left(\sum_{k=1}^{n} X_k\right) \to 0 \qquad (5.1.5)$$

as $n \to \infty$. Condition formula (5.1.5) is often called Markov condition. As an application of Markov condition, we have the following Poisson WLLN.

Theorem 5.1.2 (Poisson WLLN)

In a sequence of independent experiments, suppose that $P(A) = p_k$ in the kth experiment. Let N_n be the number of A occurring in the first n experiments. For any $\varepsilon > 0$, we have

$$\lim_{n\to\infty} P\left\{\left|\frac{N_n}{n} - \frac{p_1 + p_2 + \cdots + p_n}{n}\right| \geqslant \varepsilon\right\} = 0.$$

Proof.

As in Corollary 5.1.1, let

$$\xi_i = \begin{cases} 1, & \text{if the event A occurs in the ith experiment;} \\ 0, & \text{otherwise.} \end{cases}$$

Then

$$E(\xi_k) = p_k, \quad \mathrm{Var}(\xi_k) = p_k(1 - p_k) \leqslant \frac{1}{4}.$$

This, and Markov condition formula (5.1.5), follows the result.

The SLLN is probably the best-known result in probability theory. It states that the average of a sequence of i.i.d. random variables with finite mean, converge to the mean value with probability one. We shall show the SLLN and omit its proof for our knowledge limitation.

Theorem 5.1.3 (SLLN)

Let $X_1, X_2, \cdots$ be a sequence of i. i. d. random variables with mean $E(X_i)=\mu$. Then, with probability one,

$$\frac{X_1+X_2+\cdots+X_n}{n} \to \mu \text{ as } n\to\infty. \tag{5.1.6}$$

By Theorem 5.1.3, it says that $n_h \to 1/2$ with probability one.

The SLLN was originally proved, in the special case of Bernoulli random variables, by the French mathematician Borel. The general form of the SLLN presented in Theorem 5.1.3 was proved by the Russian mathematician A. N. Kolmogorov.

Many students are initially confused about the difference between the WLLN and SLLN. The WLLN states that for any given large value n_0, $(X_1+X_2+\cdots+X_{n_0})/n_0$ is likely to be near the mean value μ. However, it does not tell us that $(X_1+X_2+\cdots+X_n)/n$ is bounded to stay near μ for all values of n larger than n_0. Thus it leaves open the possibility that large values of $|(X_1+X_2+\cdots+X_n)/n-\mu|$ can occur infinitely often (though at infrequent intervals). The SLLN shows that this can not occur. Specially it tells that with probability one, for any positive $\varepsilon>0$,

$$\left|\frac{\sum_{k=1}^{n} X_k}{n}-\mu\right|$$

will be greater than ε only a finite number of times.

5.2 the Central Limit Theorem

The central limit theorem (CLT) is one of the most remarkable results in probability theory. It says that the sum of a large number of independent random variables has a distribution which is approximately normal. So, it not only provides a simple method for computing approximate probabilities for sums of independent random variables, but it can also help us explain the remarkable fact that the empirical frequencies of so many natural populations exhibit bell-shaped normal curves.

Before presenting the CLT, we first show the definition of convergence in distribution.

Definition 5.2.1

Given a sequence of random variables $\{X_n, n=1,2,\cdots,\}$, and suppose that X_n has the distribution $F_n, n=1,2,\cdots$. If, at all continuous points x of $F(x)$,

$$\lim_{n\to\infty} F_n(x)=F(x),$$

then X_n is said to converge to X in distribution.

Below it is the simplest form of the CLT. Generally its proof is based on theory of the generating function which is not discussed in this book, and omitted.

Theorem 5.2.1

Let $X_1, X_2, \cdots$ be a sequence of i. i. d. random variables having the same mean μ and variance σ^2. Then the distribution of

$$\frac{\sum_{k=1}^{n} X_k - n\mu}{\sigma\sqrt{n}}$$

tends to the standard normal distribution as $n\to\infty$. That is, for $-\infty<a<+\infty$,

$$\lim_{n\to\infty} P\left\{\frac{\sum_{k=1}^{n} X_k - n\mu}{\sigma\sqrt{n}} \leqslant a\right\} = \frac{1}{\sqrt{2\pi}}\int_{-\infty}^{a} e^{-\frac{x^2}{2}}\,dx.$$

With Definition 5.2.1, Theorem 5.2.1 says that the random variable

$$\frac{\sum_{k=1}^{n} X_k - n\mu}{\sigma\sqrt{n}}$$

converges to a standard normal random variable in distribution.

The first version of the CLT was proved by DeMoiver around 1733 for the special case where the X_i are Bernoulli random variables with $p=1/2$: $B\left(1,\frac{1}{2}\right)$. That is, for

$$n_h = \frac{1}{n}\sum_{i=1}^{n} X_i,$$

defined in above section,

$$P\left\{\frac{n_h - n/2}{\sqrt{n/2}} \leqslant a\right\} = P\left\{\frac{\sum_{k=1}^{n} X_k - n/2}{\sqrt{n/2}} \leqslant a\right\} = \frac{1}{\sqrt{2\pi}}\int_{-\infty}^{a} \mathrm{e}^{-\frac{x^2}{2}}\,\mathrm{d}x.$$

Laplace also discovered the more general form of CLT given in Theorem 5.2.1. However his proof was not completely rigorous and, in fact, can not easily be made rigorous. A truly rigorous proof of the CLT was first presented by the Russian mathematician Liapounoff in the period 1901—1902.

Example 5.2.1

Let X_i, $i = 1, 2, \cdots, 10$, be i. i. d. random variables with uniform distribution on $(0,1)$. Find an approximation to

$$P\left\{\sum_{i=1}^{10} X_i > 6\right\}.$$

Solution.

Because $X_i \sim U(0,1), i=1,2,\cdots,10$, we have $E(X_i)=1/2$, $\mathrm{Var}(X_i)=1/12$. By CLT,

$$P\left\{\sum_{i=1}^{10} X_i > 6\right\} = P\left\{\frac{\sum_{i=1}^{10} X_i - 5}{\sqrt{10\left(\frac{1}{12}\right)}} > \frac{6-5}{\sqrt{10\left(\frac{1}{12}\right)}}\right\}$$

$$\approx 1 - \Phi(\sqrt{1.2})$$

$$\approx 1.367.$$

Hence only 14 percent of the time will $\sum_{i=1}^{10} X_i$ be greater than 6.

Example 5.2.2

Suppose a fair coin is tossed 900 times. We shall determine the probability of obtaining more than 495 heads.

Solution.

For $i=1,\cdots,900$, define

$$X_i=\begin{cases}1, & \text{if a head is obtained on the } i\text{th toss;}\\ 0, & \text{otherwise.}\end{cases}$$

Then $E(X_i)=1/2$ and $\mathrm{Var}(X_i)=1/4$. Let the total number of heads $H=\sum_{i=1}^{900}X_i$. So,

$$\begin{aligned}P\{H>495\}&=P\left\{\frac{H-450}{15}>\frac{495-450}{15}\right\}\\&\approx 1-\Phi(3)\\&\approx 0.001\,3.\end{aligned}$$

Example 5.2.3

The number of students who enroll in a probability course is a Poisson random variable with mean 100. The professor in charge of the course has decided that if the number enrolling is 120 or more she will teach the course in two separate classes, whereas if fewer than 120 students enroll she will teach all of the students together in a single class. What is the probability that the professor will have to teach two classes?

Solution.

The exact solution

$$\mathrm{e}^{-100}\sum_{i=120}^{\infty}\frac{100^i}{i!}$$

does not readily yield a numerical answer. But, by recalling that a Poisson random variable with mean 100 is the sum of 100 independent Poisson random variables each with mean 1, we can make use of the CLT to get an approximate solution. Let X be the number of students that enroll in this course, we have

$$
\begin{aligned}
P\{X \geqslant 120\} &= P\{X \geqslant 119.05\} \\
&= P\left\{\frac{X-100}{\sqrt{100}} \geqslant \frac{119.5-100}{\sqrt{100}}\right\} \\
&\approx 1-\Phi(1.95) \\
&\approx 0.0256,
\end{aligned}
$$

where we use the fact the variance of Poisson random variable is the same as the mean.

We end this chapter by presenting a CLT for independent and not identically distributed random variables. As before we only show the result and omit its proof.

Theorem 5.2.2

Let $X_1, X_2, \cdots$ be a sequence of independent random variables having respective means $\mu_i = E(X_i)$ and variances $\mathrm{Var}(X_i) = \sigma_i^2$. If (i) the X_i are uniformly bounded, that is, if for some L, $P\{|X_i| < L\} = 1$ for all i, and (ii) $\sum_{i=1}^{\infty} \sigma_i^2 = \infty$, then

$$
\lim_{n\to\infty} P\left\{\frac{\sum_{i=1}^{n}(X_i - \mu_i)}{\sqrt{\sum_{i=1}^{n}\sigma_i^2}} \leqslant a\right\} = \frac{1}{\sqrt{2\pi}}\int_{-\infty}^{a} \mathrm{e}^{-\frac{x^2}{2}}\,\mathrm{d}x.
$$

Exercises

5.1 The CLT is a refinement of the LLN, why?

5.2 Suppose that $\{X_k, k=1,2,\cdots\}$ is a random variable sequence such that, for all $k, l = 1, 2, \cdots$, (1) $E(X_k) = \mu$; (2) $\mathrm{Var}(X_k) \leqslant C$ for some constant C; (3) X_k and X_l are correlated if $|k-l| \leqslant 1$ and uncorrelated otherwise. Prove that, for any $\varepsilon > 0$,

$$
\lim_{n\to\infty} P\left\{\left|\frac{X_1 + X_2 + \cdots + X_n}{n} - \mu\right| \geqslant \varepsilon\right\} = 0.
$$

5.3 Suppose that $\{X_k, k=1,2,\cdots\}$ is a i. i. d. sequence of normal random variables with distribution $N(0,1)$. What are the limits

$$
\lim_{n\to\infty} P\left\{\frac{X_1 + X_2 + \cdots + X_n}{n} > 0.1\right\},
$$

$$\lim_{n\to\infty} P\left\{\frac{X_1+X_2+\cdots+X_n}{n} > -0.1\right\}.$$

5.4 Suppose that $\{X_k, k=1,2,\cdots\}$ is a i. i. d. sequence of random variables uniformly distributed over [0, 1]. Approximately determine the probability

$$\lim_{n\to\infty} P\left\{\left|\frac{X_1+X_2+\cdots+X_n}{n}-\frac{1}{2}\right| > 0.1\right\}.$$

5.5 Suppose that $\{X_k, k=1,2,\cdots\}$ is a i. i. d. sequence of random variables with mean μ and variance 9 for some constant μ. Use the CLT to determine approximately the smallest value of n for which the following relation satisfies

$$P\left\{\left|\frac{X_1+X_2+\cdots+X_n}{n}-\mu\right| < 0.3\right\} \geqslant 0.95.$$

5.6 Suppose each lamp bulb made by a factory is defective with probability 0.005. Use the CLT to approximate the probability that, in 10 000 bulbs, the number of the defective bulbs is (1) between 40 and 50; (2) greater than 70.

5.7 An insurance company has 10 000 automobile policyholders. The expected yearly claim per policyholder is 240 with a standard deviation of 800. Use the CLT to approximate the probability that the total yearly claim exceeds 2.7 million.

Chapter 6 Samples and Sampling Distribution

6.1 Random Samples

Engineers and scientists are constantly exposed to collections of facts, or data, both in their professional capacities and in everyday activities. The discipline of statistics provides methods for organizing and summarizing data, and for drawing conclusions based on information contained in the data.

An investigation will typically focus on a well-defined collection of objects constituting a **population** of interest. In one study, the population might consist of all gelatin capsules of a particular type produced during a specified period. Another investigation might involve the population consisting of all **individuals** who received a B. S. in engineering during the most recent academic year. When desired information is available for all subjects in the population, we have what is called a **census**. Constraints on time, money, and other scarce resources usually make a census impractical or infeasible. Instead, a subset of the population—**a sample**—is selected in some prescribed manner. Thus we might obtain a sample of bearings from a particular production run as a basis for investigating whether bearings are conforming to manufacturing specifications, or we might select a sample of last year's engineering graduates to obtain feedback about the quality of the engineering curricula.

We are usually interested only in certain characteristics of the objects in a population: for example, the gender of an engineering graduate, the age at which the individual graduated, and so on. A characteristic may be categorical, such as gender or type of malfunction, or it may be numerical in nature. A **variable** is any characteristic whose value may change from one object to another in the population. In the following we will introduce population, random sample, observations of random sample, the size of random sample, etc.

Each individual of the population is a observation in random trials, so it is the value of a random variable X, then one population can be represented by one random variable X. Our study of the population is just about research on random variable X. The distribution function and the numerical characteristics of X is called the distribution function and numerical characteristics of population. In the future it will not distinguish between the population and the corresponding random variable, generally denoted by **population** X.

In practice, the distribution of population is unknown or it just has certain forms with unknown parameters. In statistics, people often extract part of individuals from a population to draw inferences based on the overall distribution of the data obtained. The part of individuals extracted is called **a sample** of population.

Definition 6.1.1

Suppose X is a random variable which has distribution function F, let $X_1, X_2, \cdots, X_n$ denote n independent random variables, each of which has the same but possibly unknown probability density function (often abbreviated p. d. f.) $f(x)$; that is, the joint probability density function of $X_1, X_2, \cdots, X_n$ is $f(x_1)f(x_2)\cdots f(x_n)$. The random variables $X_1, X_2, \cdots, X_n$ are then said to constitute a **simple random sample** from a distribution that has probability density function (p. d. f.) $f(x)$, and n is the **size** of this random sample. That is, the observations of a random sample are **independent and identically distributed** (often abbreviated i. i. d.), the observations $x_1, x_2, \cdots, x_n$ are sample values which are also called **independent observations** of X.

From definition we know: if $X_1, X_2, \cdots, X_n$ are independent and identically distributed (i. i. d.) samples from distribution function F, so distribution function of $X_1, X_2, \cdots, X_n$ is

$$F(x_1, x_2, \cdots, x_n) = \prod_{i=1}^{n} F(x_i).$$

Then the joint probability density of $X_1, X_2, \cdots, X_n$ is

$$f(x_1, x_2, \cdots, x_n) = \prod_{i=1}^{n} f(x_i).$$

Example 6. 1. 1

Suppose that $X_1, X_2, \cdots, X_n$ form a random sample from a Bernoulli distribution for which the parameter is p. Find the joint distribution function of random sample $X_1, X_2, \cdots, X_n$.

Solution:

The probability function of X is

$$P\{X=x\}=p^x(1-p)^{1-x}$$

Here $x=0, 1$.

So the joint probability function of $X_1, X_2, \cdots, X_n$ is

$$P\{X_1 = x_1, X_2 = x_2, \cdots, X_n = x_n\} = \prod_{i=1}^{n} p^{x_i}(1-p)^{1-x_i} = p^{\sum_{i=1}^{n} x_i}(1-p)^{n-\sum_{i=1}^{n} x_i}$$

Here $x_i=0,1(i=1,2,\cdots,n)$.

Example 6. 1. 2

Suppose that $X_1, X_2, \cdots, X_n$ form a random sample from a normal distribution for which the mean is μ and variance is σ^2. Find the joint probability density function of random sample $X_1, X_2, \cdots, X_n$.

Solution:

The probability density function of population is

$$f(x)=\frac{1}{\sqrt{2\pi}\sigma}e^{-\frac{(x-\mu)^2}{2\sigma^2}}.$$

So the joint probability density function of random samples $X_1, X_2, \cdots, X_n$ is

$$f(x_1, x_2, \cdots, x_n) = \prod_{i=1}^{n} \frac{1}{\sqrt{2\pi}\sigma} e^{-\frac{(x_i-\mu)^2}{2\sigma^2}} = \left(\frac{1}{\sqrt{2\pi}\sigma}\right)^n e^{-\frac{1}{2\sigma^2}\sum_{i=1}^{n}(x_i-\mu)^2}.$$

6.2 Statistics and Numerical Characteristics of Sample

Samples come from the population and the observations contain general information on all aspects of population, so integrating the dispersed information about the population is needed, it is necessary to process and refine the samples to reflect characteristics of population. But extracting further information of samples need to construct sample functions according to specific problems.

A statistic is a function of some observable random variables whose value can be calculated from sample data. Prior to obtaining data, there is uncertainty as to what value of any particular statistic will result. Therefore, a statistic is a random variable and will be denoted by an uppercase letter; a lowercase letter is used to represent the calculated or observed value of the statistic.

Definition 6. 2. 1

Suppose that $X_1, X_2, \cdots, X_n$ form a random sample from population X, $g(x_1, x_2, \cdots, x_n)$ is the function of random sample $X_1, X_2, \cdots, X_n$ which does not contain unknown parameter, then we call $g(x_1, x_2, \cdots, x_n)$ as a **statistic.**

Because $X_1, X_2, \cdots, X_n$ are random variables, statistic $g(x_1, x_2, \cdots, x_n)$ is the function of random variables, so the statistic is a random variable. Suppose $x_1, x_2, \cdots, x_n$ are sample values corresponding to $X_1, X_2, \cdots, X_n$, then we regard $g(x_1, x_2, \cdots, x_n)$ as the observation of $g(X_1, X_2, \cdots, X_n)$.

For example, suppose that $X_1, X_2, \cdots, X_n$ form a random sample from a normal population for which the parameters μ and σ^2 are unknown, from the definition above we know $\frac{1}{n}\sum_{i=1}^{n} X_i$, $\max(X_1, X_2, \cdots, X_n)$, $\frac{1}{n}\sum_{i=1}^{n}(X_i - \overline{X})^2$ are statistics, but $\frac{1}{n}\sum_{i=1}^{n}X_i-\mu$, $\frac{1}{\sigma^2}\sum_{i=1}^{n}(X_i-\overline{X})^2$ are not statistics.

Suppose that $X_1, X_2, \cdots, X_n$ form a random sample from population X, $x_1, x_2, \cdots, x_n$ are observations of random sample $X_1, X_2, \cdots, X_n$, in the follows we will define five principle statistics and the corresponding observations:

Sample mean

$$\overline{X}=\frac{1}{n}\sum_{i=1}^{n}X_i;$$

Sample variance

$$S^2=\frac{1}{n-1}\sum_{i=1}^{n}(X_i-\overline{X})^2=\frac{1}{n-1}(\sum_{i=1}^{n}X_i^2-n\overline{X}^2);$$

Sample standard deviation

$$S=\sqrt{S^2}=\sqrt{\frac{1}{n-1}\sum_{i=1}^{n}(X_i-\overline{X})^2};$$

The *k*th origin moment of sample

$$A_k=\frac{1}{n}\sum_{i=1}^{n}X_i^k,\quad k=1,2,\cdots;$$

The *k*th central moment of sample

$$B_k=\frac{1}{n}\sum_{i=1}^{n}(X_i-\overline{X})^k,\quad k=2,3,\cdots.$$

The corresponding observations are :

$$\overline{x}=\frac{1}{n}\sum_{i=1}^{n}x_i;$$

$$s^2=\frac{1}{n-1}\sum_{i=1}^{n}(x_i-\overline{x})^2=\frac{1}{n-1}(\sum_{i=1}^{n}x_i^2-n\overline{x}^2);$$

$$s=\sqrt{\frac{1}{n-1}\sum_{i=1}^{n}(x_i-\overline{x})^2};$$

$$a_k=\frac{1}{n}\sum_{i=1}^{n}x_i^k,\quad k=1,2,\cdots;$$

$$b_k = \frac{1}{n}\sum_{i=1}^{n}(x_i - \bar{x})^k, \quad k = 2,3,\cdots.$$

Example 6.2.1

Randomly select 10 students' scores from English final examination in one class, the scores were

100, 85, 70, 65, 90, 95, 63, 50, 77, 86

(1) Find population, samples, sample values, sample size;

(2) Find sample mean, sample variance and the second origin moment of samples.

Solution:

(1) population: English final examination scores of all students in the class X.

samples: $X_1, X_2, \cdots, X_{10}$

sample values: $(x_1, x_2, \cdots, x_{10}) = (100, 85, 70, 65, 90, 95, 63, 50, 77, 86)$

sample size: $n=10$

(2) sample mean : $\bar{x} = \frac{1}{10}\sum_{i=1}^{10} x_i = 78.1$

sample variance : $s^2 = \frac{1}{10-1}\sum_{i=1}^{10}(x_i - \bar{x})^2 = 252.5$

the second origin moment of sample: $A_2 = \frac{1}{10}\sum_{i=1}^{10} x_i^2 = 6\,326.9$

Any statistic, being a random variable, has a probability distribution. In particular, the sample mean $\overline{X}$ has a probability distribution. Suppose, for example, that $n=2$ components are randomly selected. Possible values for the sample mean are 0(if $X_1 = X_2 = 0$), and so on, from which other probabilities such as $P(1 \leqslant \overline{X} \leqslant 3)$ and $P(\overline{X} \geqslant 2.5)$ can be calculated. Similarly, if for a sample of size $n=2$, the only possible values of the sample variance are 0, 12.5, and 50 (which is the case if X_1 and X_2 can each take on only the values 40, 45and 50), then the probability distribution of S^2 gives $P(S^2=0)$, $P(S^2=12.5)$ and $P(S^2=50)$. The probability distribution of a statistic is sometimes referred to as its **sampling distribution** to emphasize that it describes how the statistic varies in value across all samples that might be selected. In the next section, we will introduce sampling distribution in details and show you some important statistics and their properties.

6.3 Sampling Distribution

Sampling distributions are important in statistics because they provide a major simplification route to statistical inference. More specifically, they allow analytical considerations to be based on the sampling distribution of a statistic, rather than on the joint probability distribution of all the individual sample values.

The sampling distribution of a statistic is the distribution of that statistic, considered as a random variable, when derived from a random sample of size n. It may be considered as the distribution of the statistic for all possible samples from the same population of a given size. The sampling distribution depends on the underlying distribution of the population, the statistic being considered, the sampling procedure employed, and the sample size used. There is often considerable interest in whether the sampling distribution can be approximated by an asymptotic distribution, which corresponds to the limiting case as $n \to \infty$.

For example, assume we repeatedly take samples of a given size from a normal population with mean μ and variance σ^2, and calculate the arithmetic mean $\bar{x}$ for each sample — this statistic is called the sample mean. Each sample has its own average value, and the distribution of these averages is called the "sampling distribution of the sample mean". This distribution is a normal population $N\left(\mu, \frac{\sigma^2}{n}\right)$ (n is the sample size) since the underlying population is normal, although sampling distributions may also often be close to normal even when the population distribution is not (see central limit theorem). An alternative to the sample mean is the sample median. When calculated from the same population, it has a different sampling distribution to that of the mean and is generally not normal (but it may be close for large sample sizes).

The mean of a sample from a population having a normal distribution is an example of a simple statistic taken from one of the simplest statistical populations. For other statistics and other populations the formulas are more complicated. The distribution of the statistic needs to be known when using it to do statistical inference. The sampling distribution is determined when distribution function is given, however, it is difficult to find precise distribution of the statistic. Then in the following some important statistics which come from normal distribution and their properties will be introduced in details.

1. Chi-square Distribution

The family of chi-square (χ^2) distributions is a subcollection of the family of gamma distributions. These special gamma distributions arise as sampling distributions of variance estimators based on random samples from a normal distribution.

In this section we will introduce and discuss a particular type of gamma distribution known as the chi-square distribution. This distribution, which is closely related to random samples from a normal distribution, is widely applied in the field of statistics; in the remainder of this book we shall see how it is applied in many important problems of statistical inference. In this section we shall present the definition of the χ^2 distribution and some of its basic mathematical properties.

Definitions 6.3.1

Suppose that $X_1, X_2, \cdots, X_n$ form a random sample from a standard normal distribution $N(0,1)$, then we call the statistic

$$\chi^2 = X_1^2 + X_2^2 + \cdots + X_n^2$$

as Chi-square distribution with n degrees of freedom, which is denoted by

$\chi^2 \sim \chi^2(n)$. Here n indicates the number of independent random variables.

The probability density function (p. d. f.) of chi-square (χ^2) distribution is

$$f(y)=\begin{cases}\dfrac{1}{2^{n/2}\Gamma(n/2)}y^{n/2-1}e^{-y/2}, & y>0;\\ 0, & \text{otherwise.}\end{cases} \tag{6. 3. 1}$$

The picture of $f(y)$ is showed in Figure 6. 3. 1 as follows.

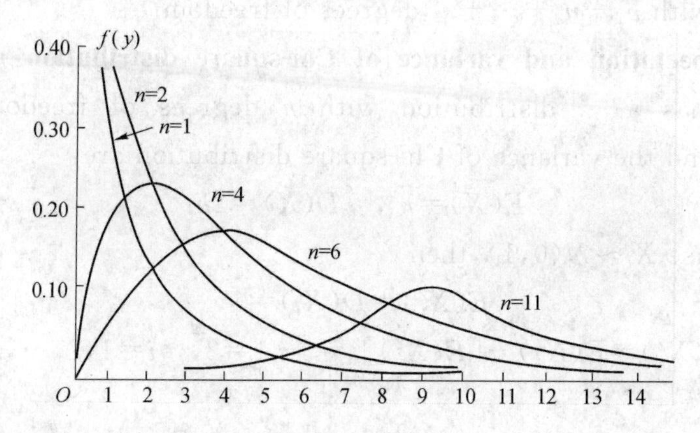

Figure 6. 3. 1

The chi-square (χ^2) distribution is a special case of Gamma distribution. The probability density function of Gamma distribution with parameters α and β is defined as follows:

$$f(x,\alpha,\beta)=\begin{cases}\dfrac{\beta^{\alpha}}{\Gamma(\alpha)}x^{\alpha-1}e^{-\beta x}, & x>0;\\ 0, & \text{otherwise.}\end{cases}$$

For each positive integer n, the gamma distribution for which $\alpha=n/2$ and $\beta=1/2$ is called the χ^2 distribution with n degrees of freedom. For example, $\chi^2(1)$ is the distribution of $\Gamma\left(\frac{1}{2},\frac{1}{2}\right)$, because of $X_i \sim N(0,1)$, $X_i^2 \sim \chi^2(1)$, that is $X_i^2 \sim \Gamma\left(\frac{1}{2},\frac{1}{2}\right)$, $i=1,2,\cdots,n$, $X_1^2, X_2^2, \cdots, X_n^2$ are independent, so

$$\chi^2 = \sum_{i=1}^{n} X_i^2 \sim \Gamma\left(\frac{n}{2},\frac{1}{2}\right).$$

Then we get the probability density function of χ^2 distribution as

Eq. (6.3.1) shows.

Properties of the Chi-square Distribution

(1) **The additivity property of the Chi-square distribution**: If the random variables $X_1, X_2, \cdots, X_k$ are independent and if X_i has a χ^2 distribution with n_i degrees of freedom ($i=1,2,\cdots,k$), then the sum $X_1+X_2+\cdots+X_k$ has a χ^2 distribution with $n_1+n_2+\cdots+n_k$ degrees of freedom.

(2) **Expectation and variance of Chi-square distribution**: If a random variable X has a χ^2 distribution with n degrees of freedom, then the expectation and the variance of Chi-square distribution are:

$$E(X)=n, \quad D(X)=2n.$$

In fact, because $X_i \sim N(0,1)$, then

$$E(X_i^2)=D(X_i)=1,$$

$$D(X_i^2)=E(X_i^4)-[E(X_i^2)]^2=3-1=2, \quad i=1,2,\cdots,n.$$

So that

$$E(\chi^2)=E\left(\sum_{i=1}^{n} X_i^2\right)=\sum_{i=1}^{n} E(X_i^2)=n,$$

$$D(\chi^2)=D\left(\sum_{i=1}^{n} X_i^2\right)=\sum_{i=1}^{n} D(X_i^2)=2n.$$

(3) **Upper α quantile of Chi-square distribution**: For a given $\alpha, 0<\alpha<1$, if the point $\chi_\alpha^2(n)$ satisfies the condition

$$P\{\chi^2 > \chi_\alpha^2(n)\}=\int_{\chi_\alpha^2(n)}^{+\infty} f(y)\,\mathrm{d}y=\alpha,$$

then we call the point $\chi_\alpha^2(n)$ as upper αquantile of $\chi^2(n)$ distribution. As is shown in Figure 6.3.2.

For example, when $\alpha=0.1, n=25$, from distribution table we find $\chi_{0.1}^2(25)=34.382$, R. A. Fisher had proved that when n is large enough ($n>40$),

$$\chi_\alpha^2(n) \approx \frac{1}{2}\left(z_\alpha+\sqrt{2n-1}\right)^2$$

Where z_α is upper α quantile of standard normal distribution. Then we can calculate familiar values with the formula above when $n>40$. For example, from the formula above we get

$$\chi^2_{0.05}(50) \approx \frac{1}{2}(1.645 + \sqrt{99})^2 = 67.221$$

(From more accurate distribution table we find $\chi^2_{0.05}(50) = 67.505$).

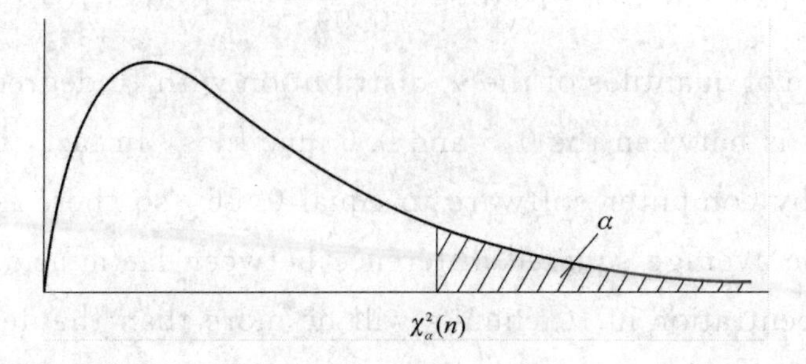

Figure 6. 3. 2

Example 6. 3. 1 Acid Concentration in Cheese

Moore and McCabe (1999) describe an experiment conducted in Australia to study the relationship between taste and the chemical composition of cheese. One chemical whose concentration can affect taste is lactic acid. Cheese manufacturers who want to establish a loyal customer base would like the taste to be about the same each time a customer purchases the cheese. The variation in concentrations of chemicals like lactic acid can lead to variation in the taste of cheese. Suppose that we model the concentration of lactic acid in several chunks of cheese as independent normal random variables with mean μ and variance σ^2. We are interested in how much these concentrations differ from the value μ. Let $X_1, \cdots, X_k$ be the concentrations in k chunks, and let $Z_i = (X_i - \mu)/\sigma$. Then

$$Y = \frac{1}{k}\sum_{i=1}^{k} |X_i - \mu|^2 = \frac{\sigma^2}{k}\sum_{i=1}^{k} Z_i^2$$

is one measure of how much the k concentrations differ from μ. Suppose that a difference of u or more in lactic acid concentration is enough to cause a noticeable difference in taste. We might then wish to calculate $P(Y \leqslant u^2)$. According to definition 6. 3. 1, the distribution of $W = kY/\sigma^2$ is χ^2 with k degrees of freedom. Hence $P(Y \leqslant u^2) = P(W \leqslant ku^2/\sigma^2)$.

For example, suppose that $\sigma^2 = 0.09$, and we are interested in $k = 10$

cheese chunks. Furthermore, suppose that $u=0.3$ is the critical difference of interest. We can write

$$P(Y\leqslant 0.3^2)=P\left(W\leqslant\frac{10\times 0.09}{0.09}\right)=P(W\leqslant 10).$$

Using the table of quantiles of the χ^2 distribution with 10 degrees of freedom, we see that 10 is between the 0.5 and 0.6 quantiles. In fact, the probability can be found by computer software to equal 0.56, so there is a 44 percent chance that the average squared difference between lactic acid concentration and mean concentration in 10 chunks will be more than the desired amount. If this probability is too large, the manufacturer might wish to invest some effort in reducing the variance of lactic acid concentration.

Example 6.3.2

Suppose that $V_1, V_2, \cdots, V_6$ form a random sample from a normal population $N(2,3)$, determine the value of b such that $P\{\sum_{i=1}^{6}(V_i-2)^2\leqslant b\}=0.95$.

Solution:

Because $\frac{V_i-2}{\sqrt{3}}\sim N(0,1), i=1,2,\cdots,6$, so

$$\sum_{i=1}^{6}\left(\frac{V_i-2}{\sqrt{3}}\right)^2\sim\chi^2(6)$$

In order to get

$$0.95=P\left\{\sum_{i=1}^{6}(V_i-2)^2\leqslant b\right\}=P\left\{\sum_{i=1}^{6}\left(\frac{V_i-2}{\sqrt{3}}\right)^2\leqslant\frac{b}{3}\right\}$$

$$=1-P\left\{\chi^2(6)>\frac{b}{3}\right\}.$$

That is

$$P\left\{\chi^2(6)>\frac{b}{3}\right\}=0.05.$$

From the distribution table, we know

$$P\{\chi^2(6)>12.592\}=0.05.$$

So

$$\frac{b}{3}=12.592, \quad b=37.776.$$

Example 6.3.3

Suppose $X_1, X_2, \cdots, X_{10}$ form a random sample from a normal population $X \sim N(0, 2^2)$, let

$$Q=aX_1^2+b(X_2+X_3)^2+c(X_4+X_5+X_6)^2+d(X_7+X_8+X_9+X_{10})^2,$$

Determine the values of a, b, c and d such that Q will have a χ^2 distribution, and find the degrees of freedom of Q.

Solution:

Because $X_1, X_2, \cdots, X_{10}$ are independent and identically distributed (i. i. d.), so

$$X_1 \sim N(0,4), \quad X_2+X_3 \sim N(0,8),$$
$$X_4+X_5+X_6 \sim N(0,12), \quad X_7+X_8+X_9+X_{10} \sim N(0,16).$$

Then $\frac{1}{2}X_1$, $\frac{1}{\sqrt{8}}(X_2+X_3)$, $\frac{1}{\sqrt{12}}(X_4+X_5+X_6)$, $\frac{1}{4}(X_7+X_8+X_9+X_{10})$ are mutually independent with same distribution $N(0,1)$, from the definition of χ^2 distribution,

$$\frac{1}{4}X_1^2+\frac{1}{8}(X_2+X_3)^2+\frac{1}{12}(X_4+X_5+X_6)^2+\frac{1}{16}(X_7+X_8+X_9+X_{10})^2 \sim \chi^2(4).$$

So, when $a=\frac{1}{4}$, $b=\frac{1}{8}$, $c=\frac{1}{12}$, $d=\frac{1}{16}$, Q will be a χ^2 distribution with 4 degrees of freedom.

2. The *T* Distribution

In this section we shall introduce and discuss another distribution, called the t distribution, which is closely related to random samples from a normal distribution. The t distribution, like the χ^2 distribution, has been widely applied in important problems of statistical inference. The t distribution is also known as **Student's distribution** in honor of W. S. Gosset, who published his studies of this distribution in 1908 under the pen name "Student." The distribution is defined as follows.

Definition 6. 3. 2

Consider two independent random variables X and Y, Y has a χ^2 distribution with n degrees of freedom and X has a standard normal distribution, namely $X \sim N(0,1)$, $Y \sim \chi^2(n)$, suppose that a random variable T is defined by the equation

$$T = \frac{X}{\sqrt{Y/n}}.$$

Then the distribution of T is called the t distribution with n degrees of freedom, which is denoted by $T \sim t(n)$.

The probability density function (p. d. f.) of $t(n)$ is

$$h(t) = \frac{\Gamma[(n+1)/2]}{\sqrt{\pi n}\,\Gamma(n/2)}\left(1+\frac{t^2}{n}\right)^{-(n+1)/2}, \quad -\infty < t < +\infty.$$

The picture of $h(t)$ is showed in Figure 6. 3. 3 as follows, where n is the number of independent variables of Y.

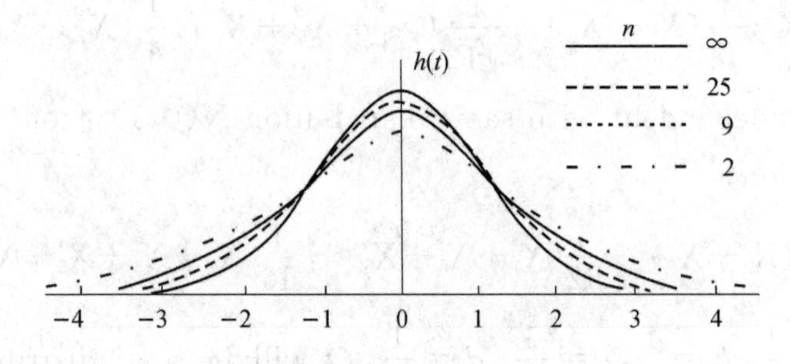

Figure 6. 3. 3

From Figure 6. 3. 3 above, we know the T distribution is symmetric and bell-shaped, like the normal distribution, but has heavier tails, meaning that it is more prone to producing values that fall far from its mean. Accordingly, the T distribution for each sample size is different, and the larger the sample, the more the distribution resembles a normal distribution, in fact, with use of characteristics of function Γ we find

$$\lim_{n\to\infty} h(t) = \frac{1}{\sqrt{2\pi}} e^{-t^2/2},$$

So, when n is large enough, T distribution is tend to standard normal *distribution* $N(0,1)$, but the difference is obvious between t distribution and standard normal distribution $N(0,1)$ when n is small.

Properties of T Distribution

Upper α quantile of T distribution: For a given $\alpha, 0<\alpha<1$, if the point $t_\alpha(n)$ satisfies the condition

$$P\{T > t_\alpha(n)\} = \int_{t_\alpha(n)}^{\infty} h(t)\mathrm{d}t = \alpha,$$

then we call the point $t_\alpha(n)$ as upper α quantile of $t(n)$ distribution. As is shown in Figure 6.3.4

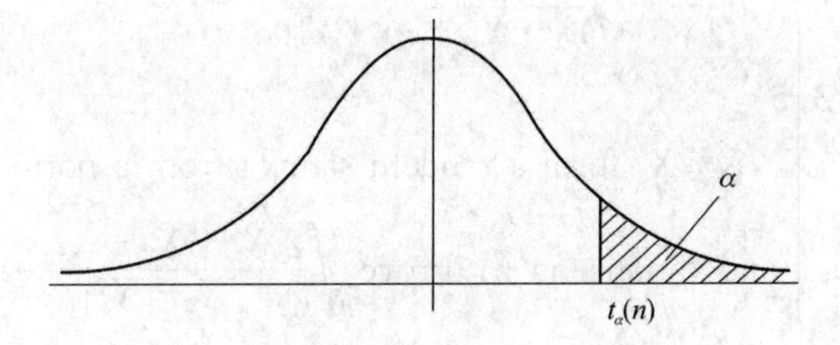

Figure 6.3.4

From the definition and the symmetric property of Figure 6.3.3 above, we know

$$t_{1-\alpha}(n) = -t_\alpha(n).$$

And when $n > 45$, for the usual value of α, we usually use approximated formula

$$t_\alpha(n) \approx z_\alpha.$$

Example 6.3.4

Suppose that $X_1, X_2, \cdots, X_9$ and $Y_1, Y_2, \cdots, Y_9$ form two samples from population X and Y respectively, which are independent and each has a normal distribution with mean 0 and variance 9. Prove $T = \dfrac{X_1+X_2+\cdots+X_9}{\sqrt{Y_1^2+Y_2^2+\cdots+Y_9^2}} \sim t(9)$.

Proof:

Let $X_i' = \dfrac{X_i}{3}$, $Y_i' = \dfrac{Y_i}{3}$, $i=1,2,\cdots,9$ then

$$X_i' \sim N(0,1), \quad Y_i' \sim N(0,1).$$

Let $X' = X_1' + X_2' + \cdots + X_9'$, $Y'^2 = Y_1'^2 + Y_2'^2 + \cdots + Y_9'^2$ then

$$X' \sim N(0,9), \quad \frac{X'}{3} \sim N(0,1), \quad Y'^2 \sim \chi^2(9).$$

So

$$T = \frac{X_1 + X_2 + \cdots + X_9}{\sqrt{Y_1^2 + Y_2^2 + \cdots + Y_9^2}} = \frac{X_1' + X_2' + \cdots + X_9'}{\sqrt{Y_1'^2 + Y_2'^2 + \cdots + Y_9'^2}} = \frac{X'}{\sqrt{Y'^2}} = \frac{X'/3}{\sqrt{Y'^2/9}}.$$

X' and Y'^2 are independent, from the definition of T distribution, we get

$$T = \frac{X_1 + X_2 + \cdots + X_9}{\sqrt{Y_1^2 + Y_2^2 + \cdots + Y_9^2}} \sim t(9).$$

Example 6.3.5

Suppose that X_1, X_2, X_3 form a random sample from a normal distribution for which the mean is 0 and variance is σ^2. Prove $\sqrt{\frac{2}{3}} \frac{X_1 + X_2 + X_3}{|X_2 - X_3|} \sim t(1)$.

Proof:

Let $Y_1 = X_2 + X_3$, $Y_2 = X_2 - X_3$, so

$$\begin{aligned} \text{cov}(Y_1, Y_2) &= E(Y_1 Y_2) - E(Y_1)E(Y_2) \\ &= E((X_2 + X_3)(X_2 - X_3)) - E(X_2 + X_3)E(X_2 - X_3) \\ &= E(X_2^2) - E(X_3^2) - 0 \\ &= \sigma^2 - \sigma^2 = 0. \end{aligned}$$

So Y_1, Y_2 are independent, with the same distribution $N(0, 2\sigma^2)$, Y_1, Y_2 and X_1 are independent respectively, then

$$X_1 + X_2 + X_3 = X_1 + Y_1 \sim N(0, 3\sigma^2).$$

So

$$\frac{X_1 + X_2 + X_3}{\sigma\sqrt{3}} \sim N(0,1), \quad \left(\frac{X_2 - X_3}{\sqrt{2}\sigma}\right)^2 \sim \chi^2(1),$$

$X_1 + X_2 + X_3$ and $X_2 - X_3$ are independent, from the definition of t distribution, we have

$$\sqrt{\frac{2}{3}} \frac{X_1 + X_2 + X_3}{|X_2 - X_3|} \sim t(1).$$

3. The F Distribution

In this section we shall introduce a probability distribution, called the F distribution, which arise in many important problems of testing hypotheses in which two or more normal distributions are to be compared on the basis of random samples from each of the distributions. We shall begin by defining the F distribution and deriving its probability density function (p. d. f.)

Definition 6. 3. 3

Consider two independent random variables U and V such that U has a χ^2 distribution with n_1 degrees of freedom and V hasa χ^2 distribution with n_2 degrees of freedom, that is $U\sim\chi^2(n_1), V\sim\chi^2(n_2)$, where n_1 and n_2 are given positive integers , then we define a new random variable F as follows:

$$F=\frac{U/n_1}{V/n_2}.$$

Then the distribution of F is called an F distribution with n_1 and n_2 degrees of freedom, which is denoted by $F\sim F(n_1, n_2)$.

The probability density function (p. d. f.) of $F(n_1, n_2)$ is

$$\psi(y)=\begin{cases}\dfrac{\Gamma[(n_1+n_2)/2](n_1/n_2)^{n_1/2}y^{(n_1/2)-1}}{\Gamma(n_1/2)\Gamma(n_2/2)[1+(n_1y/n_2)]^{(n_1+n_2)/2}}, & y>0;\\ 0, & \text{otherwise}.\end{cases}$$

The picture of $\psi(y)$ is showed in Figure 6. 3. 5 as follows.

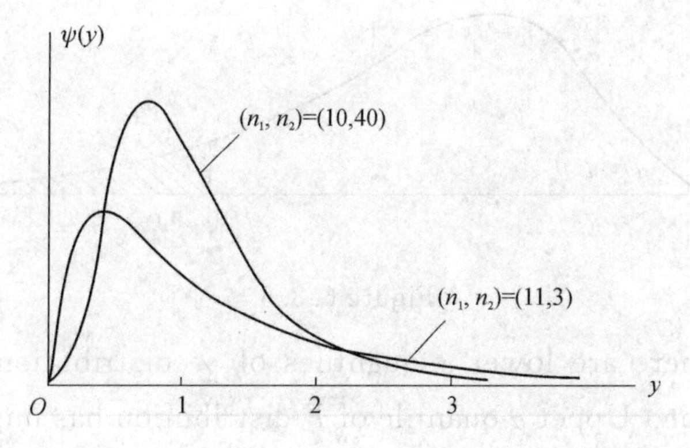

Figure 6. 3. 5

Properties of F Distribution

When we speak of an F distribution with n_1 and n_2 degrees of freedom, the order in which the numbers n_1 and n_2 are given is important, as can be seen from the definition of F distribution. When $n_1 \neq n_2$, the F distribution with n_1 and n_2 degrees of freedom and the F distribution with n_2 and n_1 degrees of freedom are two different distributions. In fact, if a random variable X has an F distribution with n_1 and n_2 degrees of freedom, then its reciprocal $1/X$ will have an F distribution with n_2 and n_1 degrees of freedom. So that if $F \sim F(n_1, n_2)$, then $\frac{1}{F} \sim F(n_2, n_1)$.

The F distribution is related to the t distribution in the following way: If a random variable X has a t distribution with n degrees of freedom, then X^2 has an F distribution with 1 and n degrees of freedom. Namely if a random variable $X \sim t(n)$, then $X^2 \sim F(1, n)$.

Upper α quantile of F distribution: For a given α, $0 < \alpha < 1$, if the point $F_\alpha(n_1, n_2)$ satisfies the condition

$$P\{F > F_\alpha(n_1, n_2)\} = \int_{F_\alpha(n_1, n_2)}^{+\infty} \psi(y)\,\mathrm{d}y = \alpha,$$

then we call the point $F_\alpha(n_1, n_2)$ as upper α quantile of $F(n_1, n_2)$ distribution. As is shown in Figure 6.3.6.

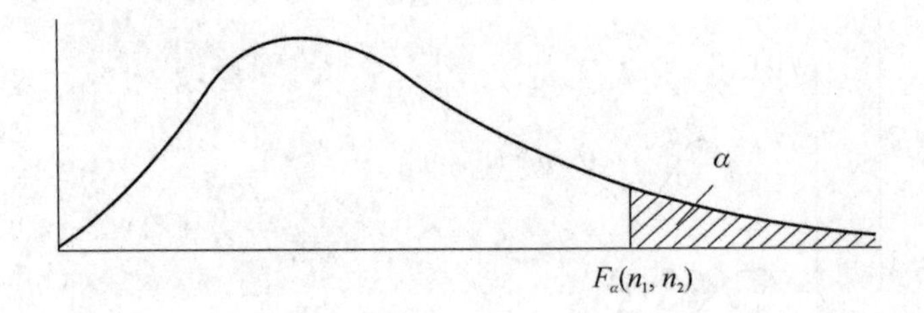

Figure 6.3.6

Similarly, there are lower α quantiles of χ^2 distribution, T distribution and F distribution. Upper α quantile of F distribution has important property as follows:

$$F_{1-\alpha}(n_1,n_2)=\frac{1}{F_\alpha(n_2,n_1)}.$$

Here we prove the formula above. If $F\sim F(n_1,n_2)$, from the definition of upper α quantile of F distribution, we have

$$1-\alpha=P\{F>F_{1-\alpha}(n_1,n_2)\}=P\left\{\frac{1}{F}<\frac{1}{F_{1-\alpha}(n_1,n_2)}\right\}$$

$$=1-P\left\{\frac{1}{F}\geqslant\frac{1}{F_{1-\alpha}(n_1,n_2)}\right\}=1-P\left\{\frac{1}{F}>\frac{1}{F_{1-\alpha}(n_1,n_2)}\right\}.$$

So

$$P\left\{\frac{1}{F}>\frac{1}{F_{1-\alpha}(n_1,n_2)}\right\}=\alpha.$$

Because $\frac{1}{F}\sim F(n_2,n_1)$, we have

$$P\left\{\frac{1}{F}>F_\alpha(n_2,n_1)\right\}=\alpha.$$

Compare the two formulas above we find

$$\frac{1}{F_{1-\alpha}(n_1,n_2)}=F_\alpha(n_2,n_1).$$

Namely

$$F_{1-\alpha}(n_1,n_2)=\frac{1}{F_\alpha(n_2,n_1)}.$$

The formula above is often used to calculate upper α quantiles of F distribution which are not listed in the table. For example,

$$F_{0.95}(12,9)=\frac{1}{F_{0.05}(9,12)}=\frac{1}{2.80}=0.357.$$

Example 6.3.6 Determining the 0.05 Quantile of an F Distribution.

Suppose that a random variable X has an F distribution with 6 and 12 degrees of freedom. We shall determine the value of x such that $P(X<x)=0.05$.

If we let $Y=1/X$, then Y will have an F distribution with 12 and 6 degrees of freedom. It can be found from the table given at the end of this book that $P(Y>4.00)=0.05$. Since the relation $Y>4.00$ is equivalent to the relation $X<0.25$, it follows that $P(X<0.25)=0.05$. Hence, $x=0.25$.

Example 6.3.7

Suppose that $X_1, X_2, \cdots, X_n$ form a random sample from a standard normal distribution. Prove $V = \left(\frac{n}{5} - 1\right)\sum_{i=1}^{5} X_i^2 / \sum_{i=6}^{n} X_i^2 \sim F(5, n-5)$.

Proof:

Because of $X_i \sim N(0,1)$, we have $\sum_{i=1}^{5} X_i^2 \sim \chi^2(5)$, $\sum_{i=6}^{n} X_i^2 \sim \chi^2(n-5)$, $\sum_{i=1}^{5} X_i^2$ and $\sum_{i=6}^{n} X_i^2$ are independent, from the definition of F distribution, we have

$$\frac{\sum_{i=1}^{5} X_i^2/5}{\sum_{i=6}^{n} X_i^2/(n-5)} \sim F(5, n-5).$$

So

$$V = \left(\frac{n}{5} - 1\right)\sum_{i=1}^{5} X_i^2 / \sum_{i=6}^{n} X_i^2 = \frac{\sum_{i=1}^{5} X_i^2/5}{\sum_{i=6}^{n} X_i^2/(n-5)} \sim F(5, n-5).$$

6.4 Distributions of Sample Mean and Sample Variance with Normal Distribution

In many practical situations, the true mean and variance of a population is not a known priori and must be computed somehow. When dealing with extremely large populations, it is not possible to count every object in the population, so the computation must be performed on a sample of the population. The sample mean (the arithmetic mean of a sample of values drawn from the population) makes a good estimator of the population mean. The sample mean (as well as sample variance) is a random variable, not a constant, since its calculated value will randomly differ depending on which members of the population are sampled, and consequently it will have its own distribution. Sample variance can also be applied to the estimation of the variance of a continuous distribution from a sample of that distribution. Then in the following we will introduce basic characteristics of sample mean and sample variance.

Suppose that $X_1, X_2, \cdots, X_n$ form a random sample from a distribution with mean μ and variance σ^2 (no matter what the distribution is, as long as the mean and the variance exist), $\overline{X}$ is sample mean and S^2 is sample variance, then

$$E(\overline{X})=\mu, D(\overline{X})=\sigma^2/n.$$

While

$$\begin{aligned} E(S^2) &= E[\frac{1}{n-1}\sum_{i=1}^{n}(X_i-\overline{X})^2] \\ &= E[\frac{1}{n-1}(\sum_{i=1}^{n}X_i^2-n\overline{X}^2)] \\ &= \frac{1}{n-1}[\sum_{i=1}^{n}E(X_i^2)-nE(\overline{X}^2)] \\ &= \frac{1}{n-1}[\sum_{i=1}^{n}(\sigma^2+\mu^2)-n(\sigma^2/n+\mu^2)] \\ &= \sigma^2. \end{aligned}$$

That is

$$E(S^2)=\sigma^2.$$

Furthermore, suppose that $X \sim N(\mu, \sigma^2)$, we find $\overline{X}=\frac{1}{n}\sum_{i=1}^{n}X_i$ also has a normal distribution. Then we have theorems as follows.

Theorem 6.4.1

Suppose that $X_1, X_2, \cdots, X_n$ form a random sample from a normal distribution with mean μ and variance σ^2, $\overline{X}$ is the sample mean and S^2 is the sample variance, then

(1) $\overline{X} \sim N(\mu, \sigma^2/n)$;

(2) $\frac{(n-1)S^2}{\sigma^2} \sim \chi^2(n-1)$;

(3) $\overline{X}$ and S^2 are independent.

Here we only give proofs of the second and the third of the Theorem 6.4.1.

Proof. Find an nth orthogonal matrix $\boldsymbol{A}=(a_{ij})$, the elements of the first row are $\frac{1}{\sqrt{n}}$, for orthogonal transformation

$$Y=\boldsymbol{A}X.$$

In which $Y=(Y_1,Y_2,\cdots,Y_n)'$, $X=(X_1,X_2,\cdots,X_n)'$, because $X_1,X_2,\cdots,X_n$ are independent and they all have a normal distribution $N(\mu,\sigma^2)$, so

$$X=(X_1,X_2,\cdots,X_n)' \sim N(\boldsymbol{\mu},\sigma^2\boldsymbol{I})$$

Where $\boldsymbol{\mu}=(\mu,\mu,\cdots,\mu)'$, $\boldsymbol{I}$ is nth unit matrix, from characteristics of normal distribution we have,

$$Y=(Y_1,Y_2,\cdots,Y_n)' \sim N(\boldsymbol{A\mu},\sigma^2\boldsymbol{AA}')$$

Because $\boldsymbol{A}$ is orthogonal matrix, so $\boldsymbol{AA}'=\boldsymbol{I}$,

$$\boldsymbol{A\mu}=\begin{bmatrix} \frac{1}{\sqrt{n}} & \frac{1}{\sqrt{n}} & \cdots & \frac{1}{\sqrt{n}} \\ * & * & \cdots & * \\ \vdots & \vdots & & \vdots \\ * & * & \cdots & * \end{bmatrix} \begin{bmatrix} \mu \\ \mu \\ \vdots \\ \mu \end{bmatrix} = \begin{bmatrix} \sqrt{n}\mu \\ 0 \\ \vdots \\ 0 \end{bmatrix}$$

So

$$Y \sim N(V,\sigma^2\boldsymbol{I})$$

Where $V=\boldsymbol{A\mu}=(\sqrt{n}\mu,0,\cdots,0)'$.

From characteristics of normal distribution we know $Y_1,Y_2,\cdots,Y_n$ are independent, $Y_1 \sim N(\sqrt{n}\mu,\sigma^2)$, $Y_i \sim N(0,\sigma^2)$, $i=2,3,\cdots,n$. Then

$$Y_1 = \frac{1}{\sqrt{n}}\sum_{i=1}^{n} X_i = \sqrt{n}\,\overline{X},$$

$$\sum_{i=2}^{n} Y_i^2 = \sum_{i=1}^{n} Y_i^2 - Y_1^2$$

$$= \sum_{i=1}^{n} X_i^2 - (\sqrt{n}\,\overline{X})^2$$

$$= \sum_{i=1}^{n} X_i^2 - n\overline{X}^2$$

$$= \sum_{i=1}^{n} (X_i - \overline{X})^2.$$

So $\overline{X} = \frac{1}{\sqrt{n}} Y_1$ and $S^2 = \frac{1}{n-1}\sum_{i=1}^{n}(X_i - \overline{X})^2 = \frac{1}{n-1}\sum_{i=2}^{n} Y_i$ are independent, and

$$\frac{(n-1)S^2}{\sigma^2} = \frac{\sum_{i=1}^{n}(X_i - \overline{X})^2}{\sigma^2} = \frac{1}{\sigma^2}\sum_{i=2}^{n} Y_i^2 = \sum_{i=2}^{n}\left(\frac{Y_i}{\sigma}\right)^2 \sim \chi^2(n-1).$$

Theorem 6. 4. 2

Suppose that $X_1, X_2, \cdots, X_n$ form a random sample from a normal distribution with mean μ and variance σ^2, $\overline{X}$ is the sample mean and S^2 is the sample variance, then

$$\frac{\overline{X}-\mu}{S/\sqrt{n}} \sim t(n-1).$$

Proof.

From Theorem 6. 4. 1 we know

$$\frac{\overline{X}-\mu}{\sigma/\sqrt{n}} \sim N(0,1), \quad \frac{(n-1)S^2}{\sigma^2} \sim \chi^2(n-1).$$

And both of them are independent, from the definition of T distribution, we have

$$\frac{\dfrac{\overline{X}-\mu}{\sigma/\sqrt{n}}}{\sqrt{\dfrac{(n-1)S^2}{\sigma^2(n-1)}}} \sim t(n-1).$$

Simplify the formula above, that is $\frac{\overline{X}-\mu}{S/\sqrt{n}} \sim t(n-1)$, then the Theorem 6. 4. 2 is followed.

Theorem 6. 4. 3

Suppose that $X_1, X_2, \cdots, X_n$ and $Y_1, Y_2, \cdots, Y_n$ form two samples from population X and Y respectively, X and Y are independent with X having a normal distribution $N(\mu_1, \sigma_1^2)$ and Y having a normal distribution $N(\mu_2, \sigma_2^2)$. Suppose $\overline{X} = \frac{1}{n_1}\sum_{i=1}^{n_1} X_i$, $\overline{Y} = \frac{1}{n_2}\sum_{i=1}^{n_2} Y_i$ are sample means respectively; $S_1^2 = \frac{1}{n_1 - 1}\sum_{i=1}^{n_1}(X_i - \overline{X})^2$, $S_2^2 = \frac{1}{n_2 - 1}\sum_{i=1}^{n_2}(Y_i - \overline{Y})^2$ are sample variances respectively, then we have

(1) $\frac{S_1^2/S_2^2}{\sigma_1^2/\sigma_2^2} \sim F(n_1 - 1, n_2 - 1)$;

(2) When $\sigma_1^2 = \sigma_2^2 = \sigma^2$,

$$\frac{(\overline{X}-\overline{Y})-(\mu_1-\mu_2)}{S_w\sqrt{\frac{1}{n_1}+\frac{1}{n_2}}} \sim t(n_1+n_2-2),$$

here

$$S_w^2 = \frac{(n_1-1)S_1^2+(n_2-1)S_2^2}{n_1+n_2-2}, \quad S_w = \sqrt{S_w^2}.$$

Proof:

(1) From Theorem 6.4.1

$$\frac{(n_1-1)S_1^2}{\sigma_1^2} \sim \chi^2(n_1-1), \quad \frac{(n_2-1)S_2^2}{\sigma_2^2} \sim \chi^2(n_2-1).$$

From the assumption, we know S_1^2, S_2^2 are independent, then from the definition of F distribution, we know

$$\frac{\frac{(n_1-1)S_1^2}{(n_1-1)\sigma_1^2}}{\frac{(n_2-1)S_2^2}{(n_2-1)\sigma_2^2}} \sim F(n_1-1, n_2-1),$$

that is

$$\frac{S_1^2/S_2^2}{\sigma_1^2/\sigma_2^2} \sim F(n_1-1, n_2-1).$$

(2) It is easy to know $\overline{X}-\overline{Y} \sim N\left(\mu_1-\mu_2, \frac{\sigma^2}{n_1}+\frac{\sigma^2}{n_2}\right)$, then

$$U=\frac{(\overline{X}-\overline{Y})-(\mu_1-\mu_2)}{\sigma\sqrt{\frac{1}{n_1}+\frac{1}{n_2}}}\sim N(0,1),$$

From the conditions which are given, we get

$$\frac{(n_1-1)S_1^2}{\sigma^2}\sim\chi^2(n_1-1),\quad \frac{(n_2-1)S_2^2}{\sigma^2}\sim\chi^2(n_2-1).$$

And they are independent, from additivity of Chi-square distribution, we know

$$V=\frac{(n_1-1)S_1^2}{\sigma^2}+\frac{(n_2-1)S_2^2}{\sigma^2}\sim\chi^2(n_1+n_2-2).$$

Because U and V are independent, from the definition of t distribution, we know

$$\frac{U}{\sqrt{V/(n_1+n_2-2)}}=\frac{(\overline{X}-\overline{Y})-(\mu_1-\mu_2)}{S_w\sqrt{\frac{1}{n_1}+\frac{1}{n_2}}}\sim t(n_1+n_2-2).$$

The several distributions and the theorems introduced in this section will play an important part in later chapters, however, we should pay attention that they are derived on the basis of population which has normal distribution.

Example 6.4.1

Suppose lives of some bulbs have a normal distribution $N(\mu,\sigma^2)$, for which μ is unknown and σ^2 is 100.

(1) Randomly select 100 bulbs, $\overline{x}$ is sample mean, find $P\{|\overline{x}-\mu|<1\}$.

(2) If $P\{|\overline{x}-\mu|<2\}\geqslant 0.95$, find the sample size n.

Solution:

(1) From the condition above, we know $\sigma/\sqrt{n}=1$, then

$$\begin{aligned}P\{|\overline{x}-\mu|<1\}&=P\{-1<\overline{x}-\mu<1\}\\&=P\left\{\frac{-1}{\sigma/\sqrt{n}}<\frac{\overline{x}-\mu}{\sigma/\sqrt{n}}<\frac{1}{\sigma/\sqrt{n}}\right\}\\&=\Phi\left(\frac{1}{\sigma/\sqrt{n}}\right)-\Phi\left(\frac{-1}{\sigma/\sqrt{n}}\right)\\&=2\Phi(1)-1=0.6826.\end{aligned}$$

(2) If $P\{|\overline{x}-\mu|<2\}\geqslant 0.95$, then

$$P\left\{\frac{-2}{\sigma/\sqrt{n}}<\frac{\overline{x}-\mu}{\sigma/\sqrt{n}}<\frac{2}{\sigma/\sqrt{n}}\right\}$$

$$=\Phi\left(\frac{2}{\sigma/\sqrt{n}}\right)-\Phi\left(\frac{-2}{\sigma/\sqrt{n}}\right)$$

$$=2\Phi\left(\frac{2}{\sigma/\sqrt{n}}\right)-1\geqslant 0.95.$$

So

$$\Phi\left(\frac{2}{\sigma/\sqrt{n}}\right)\geqslant 0.975=\Phi(1.96).$$

Then we get

$$\frac{2}{\sigma/\sqrt{n}}\geqslant 1.96,\quad n\geqslant 96.04.$$

So when $n=97$, $P\{|\overline{x}-\mu|<2\}\geqslant 0.95$.

Example 6.4.2 Suppose that the variances of two normal populations X, Y are $\sigma_1^2=12$, $\sigma_2^2=18$, respectively, randomly selecting $n_1=61$, $n_2=31$ samples from each population, two samples are independent and the sample variances are s_1^2, s_2^2, find $P\{s_1^2/s_2^2>1.16\}$.

Solution:

From Theorem 6.4.3, we know

$$\frac{s_1^2/s_2^2}{\sigma_1^2/\sigma_2^2}=\frac{s_1^2/s_2^2}{12/18}\sim F(60,30).$$

So

$$P\{s_1^2/s_2^2>1.16\}=P\left\{\frac{s_1^2/s_2^2}{\sigma_1^2/\sigma_2^2}>\frac{1.16}{12/18}\right\}=P\left\{\frac{s_1^2/s_2^2}{\sigma_1^2/\sigma_2^2}>1.74\right\}.$$

From the table we know, $F_{0.05}(60,30)=1.74$, so $P\{s_1^2/s_2^2>1.16\}=0.05$.

Exercises

6.1 Suppose that a random sample is to be taken from a normal distribution for which the value of the mean θ is unknown and the standard deviation is 2. How large a random sample must be taken in order that $E_\theta(|\overline{X}-\theta|^2)\leqslant 0.1$ for every possible value of θ ?

6.2 Randomly selecting X_1, X_2, X_3, X_4, X_5 from population $N(12,4)$,

(1) Find $P\{|\overline{X}-12|>1\}$.

(2) Find $P\{\max\{X_1, X_2, X_3, X_4, X_5\}>15\}$; $P\{\min\{X_1, X_2, X_3, X_4, X_5\}<10\}$.

6.3 Suppose that $X_1, X_2, \cdots, X_n$ form a random sample from population X which has a Poisson distribution with parameter λ.

(1) Find the joint distribution function of $X_1, X_2, \cdots, X_n$.

(2) Find $E(\overline{X})$, $D(\overline{X})$ and $E(S^2)$.

6.4 Suppose that $X_1, X_2, \cdots, X_{2n}$ form a random sample from population X with mean μ and variance σ^2, and $\overline{X}=\frac{1}{2n}\sum_{i=1}^{2n}X_i$, $Y=\sum_{i=1}^{n}(X_i+X_{n+i}-2\overline{X})^2$. Find $E(Y)$.

6.5 Randomly selecting $X_1, X_2, \cdots, X_{36}$ from population $N(52, 6.3^2)$, find the probability $P\{50.8<\overline{X}<53.8\}$.

6.6 Suppose that $X_1, X_2, \cdots, X_n$ form a random sample from a normal distribution with mean μ and variance σ^2, find the distribution of

$$\frac{n(\overline{X}-\mu)^2}{\sigma^2}.$$

6.7 Suppose that $X_1, \cdots, X_6$ form a random sample from a standard normal distribution, and let

$$Y=(X_1+X_2+X_3)^2+(X_4+X_5+X_6)^2.$$

Determine the value of c such that the random variable cY will have a χ^2 distribution.

6.8 Suppose that $X_1, \cdots, X_5$ are independent and identically distributed (i. i. d.) and each has a standard normal distribution. Determine a constant c such that the random variable

$$\frac{c(X_1+X_2)}{(X_3^2+X_4^2+X_5^2)^{1/2}}$$

will have a t distribution.

6.9 We know $X \sim t(n)$, prove $X^2 \sim F(1,n)$.

6.10 Suppose that $X_1, \cdots, X_{n+1}$ form a random sample from population $N(\mu,\sigma^2)$, $\overline{X}_n = \frac{1}{n}\sum_{i=1}^{n} X_i$, $S_n^2 = \frac{1}{n-1}\sum_{i=1}^{n}(X_i - \overline{X})^2$. Find the distribution of

$$\frac{X_{n+1}-\overline{X}_n}{S_n}\sqrt{\frac{n}{n+1}}.$$

6.11 Suppose that $X_1, X_2, \cdots, X_n$ form a random sample from a Bernoulli distribution for which the parameter is p.

(1) Find the joint distribution function of $(X_1, X_2, \cdots, X_n)$.

(2) Find the probability function of $\sum_{i=1}^{n} X_i$.

(3) Find $E(\overline{X})$, $D(\overline{X})$ and $E(S^2)$.

6.12 Suppose that $X_1, X_2, \cdots, X_{10}$ form a random sample from normal population X, $X \sim N(\mu,\sigma^2)$

(1) Find the joint probability density function of $X_1, X_2, \cdots, X_{10}$.

(2) Find the probability density function of $\overline{X}$.

6.13 Suppose that $X_1, X_2, \cdots, X_{10}$ form a random sample from a normal population $N(0, 0.3^2)$,

(1) Find the probability density function of $\sum_{i=1}^{10} X_i^2$.

(2) Find $P\{\sum_{i=1}^{10} X_i^2 > 1.44\}$.

6.14 Suppose that $X_1, X_2, \cdots, X_n$ form a random sample from population X which has an exponential distribution with parameter λ, the probability density function is

$$f(x)=\begin{cases}\lambda e^{-\lambda x}, & x \geqslant 0; \\ 0, & x<0.\end{cases}$$

Find the distribution function of sample mean $\overline{X}$.

6.15 Suppose that we randomly select 16 individuals from population

$N(\mu,\sigma^2)$, in which the parameters μ and σ^2 are unknown.

(1) Find $P\left\{\frac{S^2}{\sigma^2}\leqslant 2.041\right\}$, where S^2 is sample variance.

(2) Find $D(S^2)$.

Chapter 7 Estimation of Parameters

7.1 Point Estimation, Moment Estimation and Maximum Likehood Estimators

7.1.1 Point Estimation

Statistical inference is almost always directed toward drawing some type of conclusion about one or more parameters (population characteristics). To do so requires that an investigator obtain sample data from each of the populations under study. Conclusions can then be based on the computed values of various sample quantities. For example, let μ (a parameter) denote the true average breaking strength of wire connections used in bonding semiconductor wafers. A random sample of $n = 10$ connections might be made, and the breaking strength of each one determined, resulting in observed strengths $x_1, x_2, \cdots, x_{10}$. The sample mean breaking strength $\bar{x}$ could then be used to draw a conclusion about the value of μ. Similarly, if σ^2 is the variance of the breaking strength distribution (population variance, another parameter), the value of the sample variance s^2 can be used to infer something about σ^2.

When discussing general concepts and methods of inference, it is convenient to have a generic symbol for the parameter of interest. We will use the Greek letter θ for this purpose. The objective of point estimation is to select a single number, based on sample data, that represents a sensible value for θ. Suppose, for example, that the parameter of interest is μ, the average lifetime of calculator batteries of a certain type. A random sample of $n=3$ batteries might yield observed lifetimes (hours) $x_1=5.0, x_2=6.4, x_3=5.9$. The computed value of the sample mean lifetime is $\bar{x}=5.77$, and it is reasonable to regard 5.77 as a very plausible value of μ—our "best guess" for the value of μ based on the available sample information.

Suppose we want to estimate a parameter of a single population based on a random sample of size n. Recall from the previous chapter that before data is available, the sample observations must be considered random variables (rv',s) $X_1, X_2, \cdots, X_n$. It follows that any function of the X_i',s—that is, any statistic—such as the sample mean $\overline{X}$ or sample standard deviation S is also a random variable. The same is true if available data consists of more than one sample. For example, we can represent tensile strengths of m type 1 specimens and n type 2 specimens by $X_1, X_2, \cdots, X_m$ and $Y_1, Y_2, \cdots, Y_n$, respectively. The difference between the two sample mean strengths is $\overline{X}-\overline{Y}$, the natural statistic for making inferences about $\mu_1-\mu_2$, the difference between the population mean strengths.

Definition 7.1.1

A **point estimate** of a parameter θ is a single number that can be regarded as a sensible value for θ. A point estimate is obtained by selecting a suitable statistic and computing its value from the given sample data. The selected statistic is called the **point estimator** of θ.

In the calculator battery example just given, the estimator used to obtain the point estimate of μ was $\overline{X}$, and the point estimate of μ was 5.77. If the three observed lifetimes had instead been $x_1=5.6, x_2=4.5, x_3=6.1$, use of the estimator $\overline{X}$ would have resulted in the estimate $\bar{x}=(5.6+4.5+6.1)/3=5.4$. The symbol $\hat{\theta}$ is customarily used to denote both the estimator of θ and the point estimate resulting from a given sample. Thus $\hat{\mu}=\overline{X}$ is read as "the point estimator of μ is the sample mean $\overline{X}$." The statement "the point estimator of μ is 5.77" can be written concisely as $\hat{\mu}=5.77$. Notice that in writing $\hat{\theta}=72.5$, there is no indication of how this point estimate was obtained (what statistic was used). It is recommended that both the estimator and the resulting estimate be reported.

Example 7.1.1

An automobile manufacturer has developed a new type of bumper, which is supposed to absorb impacts with less damage than previous bumpers. The manufacturer has used this bumper in a sequence of 25 controlled crashes against a wall each at 10 mph, using one of its compact car model. Let $X=$ the number of crashes that result in no visible damage to the automobile. The parameter to be estimated is $p=$ the portion of all such crashes that result in no damage. If X is observed to be $x=15$, the most reasonable estimator and estimate are

$$\text{estimator } \hat{p}=\frac{X}{n}, \quad \text{estimate } \frac{x}{n}=\frac{15}{25}=0.60.$$

If for each parameter of interest there were only one reasonable point estimator, there would not be much to point estimation. In most problems, though, there will be more than one reasonable estimator.

Example 7.1.2

The accompanying sample consisting of $n=20$ observations on dielectric breakdown voltage of a piece of epoxy resin appeared in the article "Maximum Likelihood Estimation" in the 3-Parameter Weibull Distribution (IEEE Trans. on Dielectrics and Elec. Insul., 1996: 43-55).

24.46 25.61 26.25 26.42 26.66 27.15 27.31 27.54 27.74 27.94

27.98 28.04 28.28 28.49 28.50 28.87 29.11 29.13 29.50 30.88

The pattern in the normal probability plot given there is quite straight, so we now assume that the distribution of breakdown voltage is normal with mean value μ. Because normal distributions are symmetric, μ is also the median lifetime of the distribution. The given observations are then assumed to be the result of a random sample $X_1, X_2, \cdots, X_{20}$ from this normal distribution. Consider the following estimators and resulting estimates for μ:

A. Estimator $= \overline{X}$, estimate $= \overline{x} = \sum_{i=1}^{n} x_i/n = 555.86/20 = 27.793$, $i = 1,2,\cdots,20$.

B. Estimator $= \widetilde{X}$, estimate $= \widetilde{x} = (27.94 + 27.98)/2 = 27.960$.

C. Estimator $= [\min(X_i) + \max(X_i)]/2$ = the average of the two extreme lifetimes,

estimate $= [\min(x_i) + \max(x_i)]/2 = (24.46 + 30.88)/2 = 27.670$, $i = 1,2,\cdots 20$.

D. Estimator $= \overline{X}_{tr(10)}$, the 10% trimmed mean (discard the smallest and largest 10% of the sample and then average),

estimate $= \overline{x}_{tr(10)} = \dfrac{555.86 - 24.46 - 25.61 - 29.50 - 30.88}{16} = 27.838$.

Each one of the estimators (A)-(D) uses a different measure of the center of the sample to estimate μ. Which of the estimates is closest to the true value? We cannot answer this without knowing the true value. A question that can be answered is, "Which estimator, when used on other samples of $X_i's$, will tend to produce estimates closest to the true value?" We will shortly consider this type of question.

Example 7.1.3

In the near future there will be increasing interest in developing low-cost Mg-based alloys for various casting processes. It is therefore important to have practical ways of determining various mechanical properties of such alloys. The article "On the Development of a New Approach for the Determination of Yield Strength in Mg-based Alloys" (Light Metal Age, Oct. 1998: 50-53) proposed an ultrasonic method for this purpose. Consider the following sample of observations on elastic modulus (GPa) of AZ91D alloy specimens from a die-casting process:

44.2 43.9 44.7 44.2 44.0 43.8 44.6 43.1

Assume that these observations are the result of a random sample X_1,

$X_2, \cdots, X_8$ from the population distribution elastic modulus under such circumstances. We want to estimate the population variance σ^2. A natural estimator is the sample variance:

$$\hat{\sigma}^2 = S^2 = \frac{\sum_{i=1}^{n}(X_i - \overline{X})^2}{n-1} = \frac{\sum_{i=1}^{n} X_i^2 - (\sum_{i=1}^{n} X_i)^2/n}{n-1}.$$

The corresponding estimate is

$$\hat{\sigma}^2 = s^2 = \frac{\sum_{i=1}^{8} x_i^2 - (\sum_{i=1}^{8} x_i)^2/8}{7} = \frac{15\,533.79 - (352.5)^2/8}{7}$$

$$= 0.251\,25 \approx 0.251.$$

The estimate of σ would then be $\hat{\sigma} = s = \sqrt{0.251\,25} = 0.501$.

An alternative estimator would result from using divisor n instead of $n-1$ (i. e. , the average squared deviation):

$$\hat{\sigma}^2 = \frac{\sum_{i=1}^{n}(X_i - \overline{X})^2}{n}, \quad \text{estimate} = \frac{1.758\,75}{8} = 0.220.$$

We will shortly indicate why many statisticians prefer S^2 to the estimator with divisor n.

In the best of all possible worlds, we could find an estimator $\hat{\theta}$ for which $\hat{\theta} = \theta$ always. However, $\hat{\theta}$ is a function of the sample $X_i's$, so it is a random variable. For some samples, $\hat{\theta}$ will yield a value larger than θ, whereas for other samples $\hat{\theta}$ will underestimate θ. If we write

$$\hat{\theta} = \theta + \varepsilon \text{ (}\varepsilon\text{ is the error of estimation).}$$

Then an accurate estimator would be one resulting in small estimation errors, so that estimated values will be near the true value. An estimator that has the properties of unbiasedness and minimum variance will often be accurate in this sense, which will be introduced in section7. 2.

7. 1. 2 Moment Estimation

In this section and next, we will discuss two "constructive" methods for obtaining point estimators: the method of moments and the method of maximum likelihood. By constructive we mean that the general definition of each type of estimator suggests explicitly how to obtain the estimator in any specific problem. Although maximum likelihood estimators are generally preferable to moment estimators because of certain efficiency properties, they often require significantly more computation than do moment estimators.

The basic idea of moment estimation is to equate certain sample characteristics, such as the mean, to the corresponding population expected values. Then solving these equations for unknown parameter values yields the estimators.

Definition 7.1.2

Suppose that $X_1, X_2, \cdots, X_n$ form a random sample from probability density function (p. d. f.) $f(x)$. For $k = 1, 2, 3, \ldots$, the ***k*th population moment**, or ***k*th moment of the distribution** $f(x)$, is $E(X^k)$, denoted by $u_k = E(X^k)$. The ***k*th sample moment is** $\frac{1}{n}\sum_{i=1}^{n} X_i^k$.

Thus the first population moment is $E(X) = \mu$ and the first sample moment is $\frac{1}{n}\sum_{i=1}^{n} X_i = \overline{X}$. The second population and sample moments are $E(X^2)$ and $\frac{1}{n}\sum_{i=1}^{n} X_i^2$, respectively. The population moments will be functions of any unknown parameters $\theta_1, \theta_2, \cdots$

Definition 7.1.3

Suppose that $X_1, X_2, \cdots, X_n$ form a random sample from a distribution with probability density function (p. d. f.) $f(x; \theta_1, \cdots, \theta_m)$, where $\theta_1, \cdots, \theta_m$ are parameters whose values are unknown. Then the **moment estimators** $\hat{\theta}_1, \cdots, \hat{\theta}_m$ are obtained by equating the first m sample moments to the corresponding first m population moments and solving for $\theta_1, \cdots, \theta_m$.

If, for example, $m = 2$, $E(X)$ and $E(X^2)$ will be functions of θ_1 and θ_2. Setting $E(X) = \frac{1}{n}\sum_{i=1}^{n} X_i = \overline{X}$ and $E(X^2) = \frac{1}{n}\sum_{i=1}^{n} X_i^2$ gives two equations in θ_1

and θ_2. The solution then defines the estimators. For estimating a population mean μ, the method gives $\mu=\overline{X}$, so the estimator is the sample mean.

Example 7.1.4

Suppose that $X_1, X_2, \cdots, X_n$ form a random sample from population X with unknown parameters mean μ and variance σ^2. Find the moment estimation of μ and σ^2.

Solution:

Because $\mu_1=\mu=E(X)$, $\mu_2=E(X^2)=\mu^2+\sigma^2$, then

$$\begin{cases} \mu = \overline{X}; \\ \mu^2+\sigma^2 = \dfrac{1}{n}\sum\limits_{i=1}^{n} X_i^2. \end{cases}$$

So we get the moment estimation of μ and σ^2,

$$\hat{\mu}=\overline{X},\quad \hat{\sigma}^2=\frac{1}{n}\sum_{i=1}^{n}(X_i-\overline{X})^2.$$

Example 7.1.5

Suppose that $X_1, X_2, \cdots, X_n$ form a random sample from population X. The probability density function (p. d. f.) of population X is

$$f(x,\theta)=\begin{cases} (\theta+1)x^{\theta}, & 0<x<1; \\ 0, & \text{otherwise}. \end{cases}$$

Find the moment estimation of unknown parameter θ.

Solution:

$$\overline{X}=E(X)=\int_{-\infty}^{+\infty} xf(x)\mathrm{d}x=\int_0^1 x(\theta+1)x^{\theta}\mathrm{d}x=\frac{\theta+1}{\theta+2}.$$

So that

$$\hat{\theta}=\frac{2\overline{X}-1}{1-\overline{X}}.$$

Example 7.1.6

Suppose that $X_1, X_2, \cdots, X_n$ form a random sample from a Uniform population $U(a,b)$ with unknown parameters a and b, $(a<b)$. Find the moment estimation of a, b.

Solution:

The first and second moment of population X is

$$\mu_1 = \frac{a+b}{2}, \quad \mu_2 = \frac{(a+b)^2}{4} + \frac{(b-a)^2}{12}.$$

then

$$\begin{cases} \dfrac{a+b}{2} = \overline{X}, \\ \dfrac{(a+b)^2}{4} + \dfrac{(b-a)^2}{12} = \dfrac{1}{n}\sum_{i=1}^{n} X_i^2. \end{cases}$$

So we get the moment estimation of a, b

$$\hat{a} = \overline{X} - \sqrt{3}\hat{\sigma}, \quad \hat{b} = \overline{X} + \sqrt{3}\hat{\sigma}.$$

Where $\hat{\sigma} = \sqrt{\dfrac{1}{n}\sum_{i=1}^{n}(X_i - \overline{X})^2}$.

Example 7.1.7

Suppose that $X_1, X_2, \cdots, X_n$ form a random sample from a generalized negative binomial distribution with parameters r and p. Find the moment estimation of r and p.

Since $E(X) = r(1-p)/p$ and $V(X) = r(1-p)/p^2$,

$$E(X^2) = V(X) + [E(X)]^2 = r(1-p)(r-rp+1)/p^2.$$

Equating $E(X)$ to $\overline{X}$ and $E(X^2)$ to $\dfrac{1}{n}\sum_{i=1}^{n} X_i^2$ eventually gives

$$\hat{p} = \frac{\overline{X}}{\dfrac{1}{n}\sum_{i=1}^{n} X_i^2 - \overline{X}^2}, \quad \hat{r} = \frac{\overline{X}^2}{\dfrac{1}{n}\sum_{i=1}^{n} X_i^2 - \overline{X}^2 - \overline{X}}.$$

As an illustration, Reep, Pollard, and Benjamin ("Skill and Chance in Ball Games," J. Royal Stat. Soc., 1971: 623-629) consider the negative binomial distribution as a model for the number of goals per game scored by National Hockey League teams. The data for 1966-1967 follows (420 games):

Goals	0	1	2	3	4	5	6	7	8	9	10
Frequency	29	71	82	89	65	45	24	7	4	1	3

Then,

$$\overline{x} = \sum x_i/420 = (0 \times 29 + 1 \times 71 + \cdots + 10 \times 3)/420 = 2.98.$$

And

$$\sum x_i^2/420 = (0^2 \times 29 + 1^2 \times 71 + \cdots + 10^2 \times 3)/420 = 12.40.$$

Thus,

$$\hat{p} = \frac{2.98}{12.40 - 2.98^2} = 0.85, \quad \hat{r} = \frac{2.98^2}{12.40 - 2.98^2 - 2.98} = 16.5.$$

Although r by definition must be positive, the denominator of $\hat{r}$ could be negative, indicating that the negative binomial distribution is not appropriate (or that the moment estimator is flawed).

7.1.3 Maximum Likelihood Estimators

Maximum likelihood estimation is a method for choosing estimators of parameters that avoids using prior distributions and loss functions. It chooses the estimate of θ as the value of θ that provides the largest value of the likelihood function.

In this section, we shall develop a relatively simple method of constructing an estimator without having to specify a loss function and a prior distribution. It is called the method of **maximum likelihood**, and it was introduced by R. A. Fisher in 1912. Maximum likelihood estimation can be applied in most problems, it has a strong intuitive appeal, and will often yield a reasonable estimator of θ. Furthermore, if the sample is large, the method will typically yield an excellent estimator of θ. For these reasons, the method of maximum likelihood is probably the most widely used method of estimation in statistics.

Definition 7.1.4

Suppose that the random variables $X_1, X_2, \cdots, X_n$ form a random sample from a discrete distribution or a continuous distribution for which the probability function (p. f.) or the probability density function (p. d. f.) is $f(x;\theta)$, where the parameter θ belongs to some parameter space Ω. Here, θ

can be either a real-valued parameter or a vector. For every observed vector $x = (x_1, x_2, \cdots, x_n)$ in the sample, the value of the joint probability function (p. f.) or joint probability density function (p. d. f.) will, as usual, be denoted by $\prod_{i=1}^{n} f(x_i;\theta)$, i. e.

$$L(x_1, x_2, \cdots, x_n;\theta) = \prod_{i=1}^{n} f(x_i;\theta) = f(x_1;\theta) f(x_2;\theta) \cdots f(x_n;\theta).$$

As before, when $L(x_1, x_2, \cdots, x_n;\theta)$ (often abbreviated $L(\theta)$) is regarded as a function of θ for a given vector x, it is called the **likelihood function.**

Suppose, for the moment, that the observed vector x came from a discrete distribution. If an estimate of θ must be selected, we would certainly not consider any value of $\theta \in \Omega$ for which it would be impossible to obtain the vector x that was actually observed. Furthermore, suppose that the probability $\prod_{i=1}^{n} f(x_i;\theta)$ of obtaining the actual observed vector x is very high when θ has a particular value, say, $\theta = \theta_0$, and is very small for every other value of $\theta \in \Omega$. Then we would naturally estimate the value of θ to be θ_0 (unless we had strong prior information that outweighed the evidence in the sample and pointed toward some other value). When the sample comes from a continuous distribution, it would again be natural to try to find a value of θ for which the probability density $\prod_{i=1}^{n} f(x_i;\theta)$ is large, and to use this value as an estimate of θ. For each possible observed vector x, we are led by this reasoning to consider a value of θ for which the likelihood function $\prod_{i=1}^{n} f(x_i;\theta)$ is a maximum and to use this value as an estimate of θ. This concept is formalized in the following definition.

Definition 7.1.5

For each possible observed vector x, let $\delta(x) \in \Omega$ denote a value of $\theta \in \Omega$ for which the likelihood function $\prod_{i=1}^{n} f(x_i;\theta)$ is a maximum, and let $\hat{\theta} = \delta(X)$

be the estimator of θ defined in this way. The estimator $\hat{\theta}$ is called the **maximum likelihood estimator** of θ. The expression **maximum likelihood estimator** or **maximum likelihood estimate** is abbreviated **M. L. E.**

It should be noted that in some problems, for certain observed vectors x, the maximum value of $\prod_{i=1}^{n} f(x_i;\theta)$ may not actually be attained for any point $\theta \in \Omega$. In such a case, an M. L. E. of θ does not exist. For certain other observed vectors x, the maximum value of $\prod_{i=1}^{n} f(x_i;\theta)$ may actually be attained at more than one point in the space Ω. In such a case, the M. L. E. is not uniquely defined, and any one of these points can be chosen as the estimate $\hat{\theta}$. In many practical problems, however, the M. L. E. exists and is uniquely defined.

We shall now illustrate the method of maximum likelihood and these various possibilities by considering several examples. In each example, we shall attempt to determine an M. L. E.

Example 7.1.8 Test for a Disease

Suppose that you are walking down the street and notice that the Department of Public Health is giving a free medical test for a certain disease. The test is 90 percent reliable in the following sense: If a person has the disease, there is a probability of 0.9 that the test will give a positive response; whereas, if a person does not have the disease, there is a probability of only 0.1 that the test will give a positive response. We shall let X stand for the result of the test, where $X=1$ means that the test is positive and $X=0$ means that the test is negative. Let the parameter space be $\Omega=\{0.1, 0.9\}$, where $\theta=0.1$ means that the person tested does not have the disease, and $\theta=0.9$ means that the person has the disease. This parameter space was chosen so that, given θ, X has a Bernoulli distribution with parameter θ. The likelihood function is

$$L(x,\theta)=\theta^{x}(1-\theta)^{1-x}.$$

If $x=0$ is observed, then

$$L(0,\theta)=\begin{cases}0.9, & \theta=0.1;\\ 0.1, & \theta=0.9.\end{cases}$$

Clearly, $\theta=0.1$ maximizes the likelihood when $x=0$ is observed. If $x=1$ is observed, then

$$L(1,\theta)=\begin{cases}0.1, & \theta=0.1;\\ 0.9, & \theta=0.9.\end{cases}$$

Clearly, $\theta=0.9$ maximizes the likelihood when $x=1$ is observed. Hence, we have that the M. L. E is

$$\hat{\theta}=\begin{cases}0.1, & X=0;\\ 0.9, & X=1.\end{cases}$$

Example 7.1.9 Sampling from a Bernoulli Distribution

Suppose that the random variables $X_1, X_2, \cdots, X_n$ form a random sample from a Bernoulli distribution for which the parameter θ is unknown, $0\leqslant\theta\leqslant1$. For all observed values $x_1, x_2, \cdots, x_n$, where each x_i is either 0 or 1, the likelihood function is

$$L(x_1, x_2, \cdots, x_n;\theta)=\prod_{i=1}^{n}\theta^{x_i}(1-\theta)^{1-x_i}. \qquad (7.1.1)$$

The value of θ that maximizes the likelihood function $L(x_1, x_2, \cdots, x_n;\theta)$ will be the same as the value of θ that maximizes$\ln L(x_1, x_2, \cdots, x_n;\theta)$, since ln is an increasing function. Therefore, it will be convenient to determine the M. L. E. by finding the value of θ that maximizes

$$\begin{aligned}\ln L(\theta) &= \sum_{i=1}^{n}[x_i\ln\theta+(1-x_i)\ln(1-\theta)]\\ &= \left(\sum_{i=1}^{n}x_i\right)\ln\theta+\left(n-\sum_{i=1}^{n}x_i\right)\ln(1-\theta). \qquad (7.1.2)\end{aligned}$$

Now calculate the derivative $\mathrm{d}\ln L(\theta)/\mathrm{d}\theta$, set this derivative equal to 0, and solve the resulting equation for θ. If $\sum_{i=1}^{n}x_i\notin\{0, n\}$, we find that the derivative is 0 at $\theta=\overline{x}$, and it can be verified that this value does indeed maximize $\ln L(\theta)$ and the likelihood function defined by Eq. (7.1.1). If $\sum_{i=1}^{n} x_i = 0$, then $\ln L(\theta)$ is a decreasing function of θ for all θ, hence Lachieves its maximum at $\theta=0$. Similarly, if $\sum_{i=1}^{n} x_i = n$, L is an increasing function, and it achieves its maximum at $\theta=1$. In these last two cases, note that the maximum of the likelihood occurs at $\theta=\overline{x}$. It follows therefore that the M. L. E. of θ is $\hat{\theta}=\overline{x}$.

If $X_1, X_2, \cdots, X_n$ are regarded as n Bernoulli trials, then the M. L. E. of the unknown probability of success on any given trial is simply the proportion of successes observed in the n trials.

Example 7.1.10 Sampling from a Normal Distribution

Suppose that $X_1, X_2, \cdots, X_n$ form a random sample from a normal distribution for which the mean μ is unknown and the variance σ^2 is known. For all observed values $x_1, x_2, \cdots, x_n$, the likelihood function$L(x_1, x_2, \cdots, x_n; \mu)$ will be

$$L(x_1, x_2, \cdots, x_n; \mu) = \frac{1}{(2\pi\sigma^2)^{n/2}} \exp\left[-\frac{1}{2\sigma^2}\sum_{i=1}^{n}(x_i-\mu)^2\right] \quad (7.1.3)$$

It can be seen from Eq. (7.1.3) that $L(x_1, x_2, \cdots, x_n; \mu)$ will be maximized by the value μ that minimizes

$$Q(\mu) = \sum_{i=1}^{n}(x_i-\mu)^2 = \sum_{i=1}^{n} x_i^2 - 2\mu\sum_{i=1}^{n} x_i + n\mu^2.$$

If we now calculate the derivative $\mathrm{d}Q(\mu)/\mathrm{d}\mu$, set this derivative equal to 0, and solve the resulting equation for μ, we find that $\mu=\overline{x}$. It follows, therefore, that the M. L. E. of μ is $\hat{\mu}=\overline{X}$.

It can be seen that the estimator $\hat{\mu}$ is not affected by the value of the variance σ^2, which we assumed was known. The M. L. E. of the unknown mean μ is simply the sample mean $\overline{X}$, regardless of the value of σ^2. We shall see this again in the next example, in which both μ and σ^2 must be estimated.

Example 7.1.11 Sampling from a Normal Distribution with Unknown Variance

Suppose again that $X_1, X_2, \cdots, X_n$ form a random sample from a normal distribution, but suppose now that both the mean μ and the variance σ^2 are unknown. The parameter is then $\theta=(\mu,\sigma^2)$. For all observed values $x_1, x_2, \cdots, x_n$, the likelihood function $L(x_1, x_2, \cdots, x_n; \mu, \sigma^2)$ will again be given by the right side of Eq (7.1.3). This function must now be maximized over all possible values of μ and σ^2, where $-\infty<\mu<\infty$ and $\sigma^2>0$. Instead of maximizing the likelihood function $L(x_1, x_2, \cdots, x_n; \mu, \sigma^2)$ directly, it is again easier to maximize $L(x_1, x_2, \cdots, x_n; \mu, \sigma^2)$. We have

$$\ln L(\theta) = \ln \prod_{i=1}^{n} f(x_i; \mu, \sigma^2)$$

$$= -\frac{n}{2}\ln(2\pi) - \frac{n}{2}\ln \sigma^2 - \frac{1}{2\sigma^2}\sum_{i=1}^{n}(x_i-\mu)^2. \quad (7.1.4)$$

We shall find the value of $\theta=(\mu,\sigma^2)$ for which $\ln L(\theta)$ is maximum in three stages. First, for each fixed σ^2, we shall find the value $\hat{\mu}(\sigma^2)$ that maximizes the right side of Eq. (7.1.4). Second, we shall find the value $\hat{\sigma}^2$ of σ^2 that maximizes $\ln L(\theta')$ when $\theta'=(\hat{\mu}(\sigma^2),\sigma^2)$. Finally, the M. L. E. of θ will be the random vector whose observed value is $(\hat{\mu}(\hat{\sigma}^2),\hat{\sigma}^2)$. The first stage has already been solved in Example 7.1.10. There we obtained $\hat{\mu}(\sigma^2)=\bar{x}$. For the second stage, we set $\theta'=(\bar{x},\sigma^2)$ and maximize

$$\ln L(\theta') = -\frac{n}{2}\log(2\pi) - \frac{n}{2}\log \sigma^2 - \frac{1}{2\sigma^2}\sum_{i=1}^{n}(x_i-\bar{x})^2. \quad (7.1.5)$$

This can be maximized by setting its derivative with respect to σ^2 equal to 0 and solving for σ^2. The derivative is

$$\frac{\mathrm{d}\ln L(\theta')}{\mathrm{d}\sigma^2} = -\frac{n}{2\sigma^2} + \frac{1}{2(\sigma^2)^2}\sum_{i=1}^{n}(x_i-\bar{x})^2.$$

Setting this to 0 yields

$$\sigma^2 = \frac{1}{n}\sum_{i=1}^{n}(x_i-\bar{x})^2. \quad (7.1.6)$$

The second derivative of Eq. (7.1.5) is negative at the value of σ^2 in Eq. (7.1.6), so we have found the maximum. Therefore, the M. L. E. of $\theta=(\mu,\sigma^2)$ is

$$\hat{\theta} = (\hat{\mu}, \hat{\sigma^2}) = (\overline{X}, \frac{1}{n}\sum_{i=1}^{n}(X_i - \overline{X})^2). \tag{7.1.7}$$

Recall that the first coordinate of the M. L. E. in Eq. (7.1.7) is called the sample mean of the data. Likewise, we call the second coordinate of this M. L. E. the sample variance. It is not difficult to see that the observed value of the sample variance is the variance of a distribution that assigns probability $1/n$ to each of the n observed values $x_1, \cdots, x_n$ in the sample.

Example 7.1.12 Sampling from a Uniform Distribution

Suppose that $X_1, X_2, \cdots, X_n$ form a random sample from a uniform distribution on the interval $[0, \theta]$, where the value of the parameter θ is unknown, $\theta > 0$. The probability density function (p. d. f.) $f(x_i; \theta)$ of each observation has the following form:

$$f(x_i;\theta) = \begin{cases} \frac{1}{\theta}, & 0 \leqslant x_i \leqslant \theta; \\ 0, & \text{otherwise}. \end{cases} \tag{7.1.8}$$

Therefore, the joint probability density function (p. d. f.) $\prod_{i=1}^{n} f(x_i;\theta)$ of $X_1, X_2, \cdots, X_n$ has the form

$$L(x_i;\theta) = \prod_{i=1}^{n} f(x_i;\theta) = \begin{cases} \frac{1}{\theta^n}, & 0 \leqslant x_i \leqslant \theta (i = 1, \cdots, n); \\ 0, & \text{otherwise}. \end{cases} \tag{7.1.9}$$

It can be seen from Eq. (7.1.9) that the M. L. E. of θ must be a value of θ for which $\theta \geqslant x_i$ for $i = 1, \cdots, n$ and that maximizes $1/\theta^n$ among all such values. Since $1/\theta^n$ is a decreasing function of θ, the estimate will be the smallest value of θ such that $\theta \geqslant x_i$ for $i = 1, \cdots, n$. Since this value is $\theta = \max(x_1, x_2, \cdots, x_n)$, the M. L. E. of θ is

$$\hat{\theta} = \max(X_1, X_2, \cdots, X_n).$$

Example 7.1.13

Suppose that $X_1, X_2, \cdots, X_n$ form a random sample from a Weibull probability density function (p. d. f.)

$$f(x;\alpha,\beta)=\begin{cases}\dfrac{\alpha}{\beta^{\alpha}}x^{\alpha-1}\mathrm{e}^{-(x/\beta)^{\alpha}}, & x\geqslant 0;\\ 0, & \text{otherwise.}\end{cases}$$

Writing the likelihood and ln (likelihood), then setting both

$$(\partial/\partial\alpha)[\ln(f)]=0$$

and

$$(\partial/\partial\beta)[\ln(f)]=0$$

yields the equations

$$\alpha=\left[\frac{\sum_{i=1}^{n}x_i^{\alpha}\cdot\ln(x_i)}{\sum x_i^{\alpha}}-\frac{\sum_{i=1}^{n}\ln(x_i)}{n}\right]^{-1},\quad \beta=\left(\frac{\sum_{i=1}^{n}x_i^{\alpha}}{n}\right)^{1/\alpha}.$$

These two equations cannot be solved explicitly to give general formulas for the M. L. E. $\hat{\alpha}$ and $\hat{\beta}$. Instead, for each sample $x_1,x_2,\cdots,x_n$, the equations must be solved using an iterative numerical procedure. Even moment estimators of α and β are somewhat complicated.

7.2 the Evaluation Criteria of Estimators

For the same parameter, different estimation methods may lead to different estimators, the same estimation methods may also lead to different estimators, that is to say, the same parameter may have multiple estimators, and in principle, any statistics can be estimators of unknown parameters, but which one is best to choose? Then it relates to evaluation criteria of estimators, evaluation criteria of estimators are: whether the system deviates or not, the size of volatility, whether it is more accurate accompanied by an increase in the sample size. Namely, they are unbiased estimators, efficiency of estimator, consistency of estimator.

7.2.1 Unbiased Estimator

Suppose we have two measuring instruments; one instrument has been accurately calibrated, but the other systematically gives readings smaller than the true value being measured. When each instrument is used repeatedly on the same object, because of measurement error, the observed measurements will not be identical. However, the measurements produced by the first instrument will be distributed about the true value in such a way that on average this instrument measures what it purports to measure, so it is called an unbiased instrument. The second instrument yields observations that have a systematic error component or bias.

Definition 7.2.1

Suppose that $X_1, X_2, \cdots, X_n$ form a random sample from population X with parameter θ to be estimated, $\theta \in \Theta$ and Θ stands for ranges of parameter θ. If expectation $E(\hat{\theta})$ of estimator $\hat{\theta}=\hat{\theta}(X_1, X_2, \cdots, X_n)$ exists, and for any $\theta \in \Theta$,

$$E(\hat{\theta})=\theta.$$

Then we call $\hat{\theta}$ as the unbiased estimator of θ.

So the point estimator $\hat{\theta}$ is said to be an unbiased estimator of θ if $E(\hat{\theta})=\theta$ for every possible value of θ. If $\hat{\theta}$ is not unbiased, the difference $E(\hat{\theta})-\theta$ is called the bias of $\hat{\theta}$.

Thus, $\hat{\theta}$ is unbiased if its probability (i. e., sampling) distribution is always "centered" at the true value of the parameter. Figure 7.2.1 pictures the distributions of several biased and unbiased estimators. Note that "centered" here means that the expected value, not the median, of the distribution of $\hat{\theta}$ is equal to θ. Figure 7.2.1 shows the pdf's of a biased estimator $\hat{\theta}_1$ and an unbiased estimator $\hat{\theta}_2$ for a parameter θ.

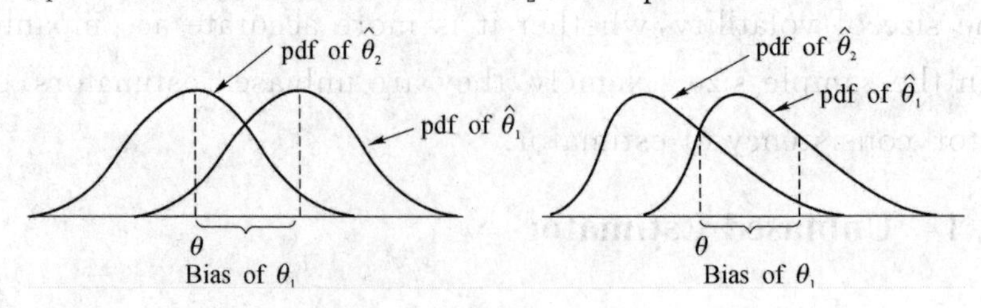

Figure 7.2.1

Example 7.2.1

Suppose that $X_1, X_2, \cdots, X_n$ form a random sample from population X, and kth moment $\mu_k = E(X^k)$ of population X exists. Prove that no matter what distribution the population has, the kth moment $A_k = \frac{1}{n}\sum_{i=1}^{n} X_i^k$ is unbiased estimator of μ_k.

Proof: Because $X_1, X_2, \cdots, X_n$ have same distribution, so

$$E(X_i^k) = E(X^k) = \mu_k, \quad i=1,2,\cdots,n.$$

Then

$$E(A_k) = \frac{1}{n}\sum_{i=1}^{n} E(X_i^k) = \mu_k.$$

Example 7.2.2

Suppose that $X_1, X_2, \cdots, X_n$ form a random sample from population X with probability density function (p. d. f.)

$$f(x;\theta) = \begin{cases} \frac{1}{\theta} e^{-x/\theta}, & x \geqslant 0; \\ 0, & x < 0. \end{cases}$$

In which $\theta > 0$ and θ is unknown.

(1) Prove $\overline{X}$ is unbiased estimator of θ, $\frac{1}{\overline{X}}$ is not unbiased estimator of $\frac{1}{\theta}$,

(2) Find constant C such that $C \min(X_1, X_2, \cdots, X_n)$ is unbiased estimator of θ.

Proof: (1) Because $E(X) = \theta$, $\overline{X}$ is unbiased estimator of θ, and $n\overline{X} = \sum_{i=1}^{n} X_i \sim \Gamma\left(n, \frac{1}{\theta}\right)$,

$$\begin{aligned} E\left[\frac{1}{\overline{X}}\right] &= nE\left[\frac{1}{n\overline{X}}\right] \\ &= n\int_0^{+\infty} \frac{1}{y} \cdot \frac{1}{\theta\Gamma(n)}\left(\frac{y}{\theta}\right)^{n-1} e^{-y/\theta} dy \\ &= \frac{n}{\theta\Gamma(n)}\Gamma(n-1) \\ &= \frac{n}{n-1}\frac{1}{\theta} \neq \frac{1}{\theta}. \end{aligned}$$

So $\frac{1}{\overline{X}}$ is not unbiased estimator of $\frac{1}{\theta}$.

(2) Suppose $Z=\min(X_1,X_2,\cdots,X_n)$, from distribution function of X

$$F(x)=\begin{cases}1-e^{-x/\theta}, & x\geqslant 0;\\ 0, & x<0.\end{cases}$$

Then distribution function of Z is

$$F_z(z)=\begin{cases}1-e^{-\frac{n}{\theta}z}, & z\geqslant 0;\\ 0, & z<0.\end{cases}$$

So probability density function (p. d. f.) of Z is

$$f_z(z)=\begin{cases}\frac{n}{\theta}e^{-\frac{n}{\theta}z}, & z\geqslant 0;\\ 0, & z<0.\end{cases}$$

Then

$$E(Z)=\frac{\theta}{n}.$$

If

$$E[C\min(X_1,X_2,\cdots,X_n)]=CE(Z)=C\frac{\theta}{n}=\theta,$$

then $C=n$. So when $C=n$, $C\min(X_1,X_2,\cdots,X_n)$ is unbiased estimator of θ.

Example 7.2.3

Suppose that X, the reaction time to a certain stimulus, has a uniform distribution on the interval from 0 to an unknown upper limit θ(so the density function of X is rectangular in shape with height $1/\theta$ for $0\leqslant x\leqslant\theta$). It is desired to estimate θ on the basis of a random sample $X_1, X_2, \cdots, X_n$ of reaction times, since θ is the largest possible time in the entire population of reaction times, consider as a first estimator the largest sample reaction time: $\hat{\theta}_1=\max(X_1,X_2,\cdots,X_n)$. If $n=5$ and $x_1=4.2, x_2=1.7, x_3=2.4, x_4=3.9$, $x_5=1.3$, the point estimate of θ is $\hat{\theta}_1=\max(4.2,1.7,2.4,3.9,1.3)=4.2$.

Unbiasedness implies that some samples will yield estimates that exceed θ and other samples will yield estimates smaller than θ—otherwise θ could not possibly be the center (balance point) of $\hat{\theta}_1$'s distribution. However, our proposed estimator will never overestimate θ (the largest sample value cannot exceed the largest population value) and will underestimate θ unless the largest sample value equals θ. This intuitive argument shows that $\hat{\theta}_1$ is a biased estimator. More precisely, it can be shown that

$$E(\hat{\theta}_1)=\frac{n}{n+1}\theta<\theta \text{ , (since } \frac{n}{n+1}<1).$$

The bias of $\hat{\theta}_1$ is given by $n\theta/(n+1)-\theta=-\theta/(n+1)$, which approaches 0 as n gets large.

It is easy to modify $\hat{\theta}_1$ to obtain an unbiased estimator of θ. Consider the estimator

$$\hat{\theta}_2=\frac{n+1}{n}\cdot\max(X_1,X_2,\cdots,X_n).$$

Using this estimator on the data gives the estimate $\frac{6}{5}\times 4.2=5.04$. The fact that $\frac{n+1}{n}>1$ implies that $\hat{\theta}_2$ will overestimate θ for some samples and underestimate it for others. The mean value of this estimator is

$$\begin{aligned}E(\hat{\theta}_2)&=E[\frac{n+1}{n}\max(X_1,X_2,\cdots,X_n)]\\&=\frac{n+1}{n}E[\max(X_1,X_2,\cdots,X_n)]\\&=\frac{n+1}{n}\cdot\frac{n}{n+1}\theta=\theta.\end{aligned}$$

If $\hat{\theta}_2$ is used repeatedly on different samples to estimate θ, some estimates will be too large and others will be too small, but in the long run there will be no systematic tendency to underestimate or overestimate θ.

7.2.2 Efficiency of Estimator

In practical problems, people primarily care about the unbiasedness of estimation, but there are several different unbiased estimators of a parameter, then which one is best to choose? A directed idea is to find an estimator which has the least volatility around the true value. In other words ,it is better to choose unbiased estimator with smallest variance, so we use variance to measure the performance of unbiased estimators. Then it comes to the efficiency of unbiased estimator which is defined as follows.

Definition 7.2.2

Suppose $\hat{\theta}_1=\hat{\theta}_1(X_1,X_2,\cdots,X_n)$ and $\hat{\theta}_2=\hat{\theta}_2(X_1,X_2,\cdots,X_n)$ are two estimators of θ that are both unbiased. Then, although the distribution of each estimator is centered at the true value of θ, the spreads of the distribution about the true value may be different. For random $\theta\in\Theta$, if

$$D(\hat{\theta}_1)\leqslant D(\hat{\theta}_2),$$

then we think $\hat{\theta}_1$ is more effective than $\hat{\theta}_2$.

Suppose $\hat{\theta}_1$ and $\hat{\theta}_2$ are two estimators of θ that are both unbiased. Then, although the distribution of each estimator is centered at the true value of θ, the spreads of the distribution about the true value may be different. Among all estimators of θ that are unbiased, choose the one that has minimum variance. The resulting $\hat{\theta}$ is called the **minimum variance unbiased estimator** (**MVUE**) of θ.

Figure 7.2.2 pictures the pdf's of two different unbiased estimators, with $\hat{\theta}_1$ having smaller variance than $\hat{\theta}_2$. Then $\hat{\theta}_1$ is more likely than $\hat{\theta}_2$ to produce an estimate close to the true θ. The MVUE is, in a certain sense, the mostly likely among all unbiased estimators to produce an estimate close to the true θ.

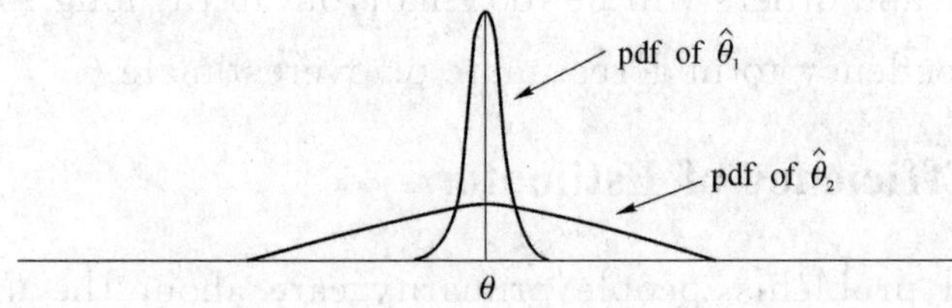

Figure 7.2.2

Example7.2.4

Suppose that $X_1, X_2, \cdots, X_n$ form a random sample from population X, $E(X)=\mu$, $D(X)=\sigma^2$, so $\hat{\mu}_1=\overline{X}, \hat{\mu}_2=X_1$ are unbiased estimators of μ, but $D(\hat{\mu}_1)=\frac{\sigma^2}{n}$, $D(\hat{\mu}_2)=\sigma^2$, when $n>1$, $\hat{\mu}_1$ is more effective than $\hat{\mu}_2$.

Example 7.2.5

We argued in example 7.2.3 that when $X_1, X_2, \cdots, X_n$ is a random sample from a uniform distribution on $[0,\theta]$, the estimator

$$\hat{\theta}_1=\frac{n+1}{n}\cdot \max(X_1, X_2, \cdots, X_n)$$

is unbiased for θ(we previously denoted this estimator by $\hat{\theta}_2$). This is not the unique unbiased estimator of θ. The expected value of a uniformly distributed r.v is just the midpoint of the interval of positive density, so $E(X_i)=\theta/2$. This implies that $E(\overline{X})=\theta/2$, from which $E(2\overline{X})=\theta$. That is, the estimator $\hat{\theta}_2=2\overline{X}$ is unbiased for θ.

If X is uniformly distributed on the interval $[A,B]$, then $V(X)=\sigma^2=(B-A)^2/12$. Thus, in our situation, $V(X_i)=\theta^2/12$, $V(\overline{X})=\theta^2/(12n)$, $V(\hat{\theta}_2)=V(2\overline{X})=4V(\overline{X})=\theta^2/3n$. Because $V(\hat{\theta}_1)=\theta^2/(n^2+2n)$, the estimator $\hat{\theta}_1$ has smaller variance than does $\hat{\theta}_2$ if $3n<n^2+2n$, as long as $n>1$, $V(\hat{\theta}_1)<V(\hat{\theta}_2)$, so $\hat{\theta}_1$ is a better estimator than $\hat{\theta}_2$. More advanced methods can be used to show that $\hat{\theta}_1$ is the MVUE of θ—every other unbiased estimator of θ has variance that exceeds $\theta^2/(n^2+2n)$.

7.2.3 Consistency of Estimator

A consistent sequence of estimators is a sequence of estimators that converge in probability to the true parameter being estimated as the sample size n tend to infinity. In other words, increasing the sample size increases the probability of the estimator being close to the population parameter.

If $n\to\infty$, the estimator $\hat{\theta}$ tend to be very close to the true parameter θ, it is said to be consistent. This notion is equivalent to convergence in probability dened below.

Definition 7.2.3

Suppose that $\hat{\theta}(X_1, X_2, \cdots, X_n)$ is the estimator of θ for random $\theta \in \Theta$, if $\hat{\theta}(X_1, X_2, \cdots, X_n)$ converges to θ in probability when $n \to \infty$, then $\hat{\theta}$ is consistent estimator of θ. Namely, for random $\varepsilon > 0$, if

$$\lim_{n \to \infty} P\{|\hat{\theta} - \theta| < \varepsilon\} = 1,$$

then we call $\hat{\theta}$ as **consistent estimator** of θ.

Consistency is a relatively weak property and is considered necessary of all reasonable estimators. This is in contrast to optimality properties such as efficiency which state that the estimator is "best".

For example, the kth moment of sample is the consistent estimator of $\mu_k = E(X^k)$ of population X, so, if $\theta = g(\mu_1, \mu_2, \cdots, \mu_k)$, where g is a continuous function, then the moment estimation $\hat{\theta} = g(\hat{\mu}_1, \hat{\mu}_2, \cdots, \hat{\mu}_k) = g(A_1, A_2, \cdots, A_k)$ of θ is a consistent estimator of θ.

7.3 Estimation of Intervals

A point estimate, because it is a single number, by itself provides no information about the precision and reliability of estimation. Consider, for example, using the statistic $\overline{X}$ to calculate a point estimate for the true average breaking strength of paper towels of a certain brand, and suppose that $\overline{x} = 93.2$. Because of sampling variability, it is virtually never the case that $\overline{x} = \mu$. The point estimate says nothing about how close it might be to μ. An alternative to reporting a single sensible value for the parameter being estimated is to calculate and report an entire interval of plausible values—an interval estimate or **confidence interval (CI)**. A confidence interval is always calculated by first selecting a confidence level (usually denoted by α), which is a measure of the degree of reliability of the interval. A confidence interval with a 95% confidence level for the true average breaking strength might have a lower limit of 91.6 and an upper limit of 94.8. Then at the 95% confidence level, any value of μ between 91.6 and 94.8 is plausible. A confidence level of 95% implies that 95% of all samples would give an interval that includes μ, or whatever other parameter is being estimated, and only 5% of all samples would yield an enormous interval. The most frequently used confidence levels are 95%, 99%, 90%. The higher the confidence level, the more strongly we believe that the value of the parameter being estimated lies within the interval (an interpretation of any particular confidence level will be given shortly).

Information about the precision of an interval estimate is conveyed by the width of the interval. If the confidence level is high and the resulting interval is quite narrow, our knowledge of the value of parameter is reasonably precise. A very wide confidence interval, however, gives the message that there is a great deal of uncertainly concerning the value of what we are estimating.

Definition 7.3.1 Confidence Intervals

As a general case, suppose that $X_1, X_2, \cdots, X_n$ form a random sample from a distribution $F(x;\theta)$ that involves a parameter θ whose value is unknown. Suppose also that two statistics $\underline{\theta}=\underline{\theta}(X_1, X_2, \cdots, X_n)$ and $\overline{\theta}=\overline{\theta}(X_1, X_2, \cdots, X_n)$ can be found such that no matter what the true value of θ may be,

$$P\{\underline{\theta}(X_1, X_2, \cdots, X_n)<\theta<\overline{\theta}(X_1, X_2, \cdots, X_n)\}=1-\alpha.$$

Where α is a fixed probability $(0<\alpha<1)$. Then it is said that the interval $(\underline{\theta}, \overline{\theta})$ is a **confidence interval** for θ with **confidence coefficient** $1-\alpha$. $\underline{\theta}$ is the lower confidence bound for θ, $\overline{\theta}$ is the upper confidence bound for θ, $1-\alpha$ is confidence coefficient.

Example 7.3.1

Suppose that $X_1, X_2, \cdots, X_n$ form a random sample from a normal *distribution* $N(\mu,\sigma^2)$ that involves parameters in which σ^2 is known and μ is unknown. Find confidence interval for μ with confidence coefficient $1-\alpha$.

Solution:

Because $\overline{X}$ is unbiased estimator of μ and

$$\frac{\overline{X}-\mu}{\sigma/\sqrt{n}}\sim N(0,1).$$

Then $\dfrac{\overline{X}-\mu}{\sigma/\sqrt{n}}$ has a standard normal distribution which does not depend on any unknown parameters. From definition of upper α quintile of standard normal distribution, so

$$P\left\{\left|\frac{\overline{X}-\mu}{\sigma/\sqrt{n}}\right|<z_{\alpha/2}\right\}=1-\alpha.$$

That is

$$P\left\{\overline{X}-\frac{\sigma}{\sqrt{n}}z_{\alpha/2}<\mu<\overline{X}+\frac{\sigma}{\sqrt{n}}z_{\alpha/2}\right\}=1-\alpha.$$

Then we get the confidence interval for μ with confidence coefficient $1-\alpha$

$$\left(\overline{X}-\frac{\sigma}{\sqrt{n}}z_{\alpha/2},\overline{X}+\frac{\sigma}{\sqrt{n}}z_{\alpha/2}\right)$$

or it is short for

$$\left(\overline{X}\pm\frac{\sigma}{\sqrt{n}}z_{\alpha/2}\right).$$

If $1-\alpha=0.95$, then $\alpha=0.05$, for $\sigma=1, n=16$, from the table we know $z_{\alpha/2}=z_{0.025}=1.96$, then we get the confidence interval with confidence coefficient 0.95,

$$\left(\overline{X}\pm\frac{1}{\sqrt{16}}\times 1.96\right)=(\overline{X}\pm 0.49).$$

So if we calculate the observed value of sample mean $\overline{x}=5.20$, we get the interval

$$(5.20\pm 0.49)=(4.71,5.69).$$

While confidence interval with confidence coefficient $1-\alpha$ is not unique. In this example we set $\alpha=0.05$, then

$$P\{-z_{0.04}<\frac{\overline{X}-\mu}{\sigma/\sqrt{n}}<z_{0.01}\}=0.95.$$

That is

$$P\{\overline{X}-\frac{\sigma}{\sqrt{n}}z_{0.01}<\mu<\overline{X}+\frac{\sigma}{\sqrt{n}}z_{0.04}\}=0.95.$$

So

$$\left(\overline{X}-\frac{\sigma}{\sqrt{n}}z_{0.01},\overline{X}+\frac{\sigma}{\sqrt{n}}z_{0.04}\right)$$

is also the confidence interval for μ with confidence coefficient 0.95, then we can compare it with $\left(\overline{X}\pm\frac{\sigma}{\sqrt{n}}z_{\alpha/2}\right)$, the interval length of $\left(\overline{X}-\frac{\sigma}{\sqrt{n}}z_{0.01},\overline{X}+\frac{\sigma}{\sqrt{n}}z_{0.04}\right)$ is $\frac{\sigma}{\sqrt{n}}(z_{0.01}+z_{0.04})=4.08\times\frac{\sigma}{\sqrt{n}}$, which is longer than the interval length of $\frac{\sigma}{\sqrt{n}}(z_{0.025}+z_{0.025})=3.92\times\frac{\sigma}{\sqrt{n}}$. Shorter confidence interval indicates the estimate is more accurate. So when n is fixed, we usually choose $\left(\overline{X}\pm\frac{\sigma}{\sqrt{n}}z_{\alpha/2}\right)$.

7.4 Interval Estimation of Normal Population Parameters

7.4.1 Interval Estimation of Single Normal Population Parameters

1. Confidence Interval for μ

When you compute a confidence interval on the mean, you compute the sample mean in order to estimate the mean of the population. Clearly, if you already knew the population mean, there would be no need for a confidence interval. However, to explain how confidence intervals are constructed, we assume that the numerical characteristics of the population are unknown. Then we will show how sample data can be used to construct a confidence interval.

Suppose that $X_1, X_2, \cdots, X_n$ form a random sample from a normal distribution with mean μ and variance σ^2, for a given $\alpha(0<\alpha<1)$, find the confidence interval for μ with confidence coefficient $1-\alpha$.

(1) When variance σ^2 is known, the confidence interval for μ with confidence coefficient $1-\alpha$ is

$$\left(\overline{X}\pm\frac{\sigma}{\sqrt{n}}z_{\alpha/2}\right).$$

Where $z_{\alpha/2}$ is upper $\alpha/2$ quantile of standard normal distribution.

(2) When variance σ^2 is unknown, we could not use the confidence interval above, because it contains unknown parameter σ, because S^2 is unbiased estimator of σ^2 and

$$t=\frac{\overline{X}-\mu}{S/\sqrt{n}}\sim t(n-1).$$

For a given α, from definition of upper α quantile of T distribution, then

$$P\left\{-t_{\alpha/2}(n-1)<\frac{\overline{X}-\mu}{S/\sqrt{n}}<t_{\alpha/2}(n-1)\right\}=1-\alpha.$$

That is

$$P\left\{\overline{X}-\frac{S}{\sqrt{n}}t_{\alpha/2}(n-1)<\mu<\overline{X}+\frac{S}{\sqrt{n}}t_{\alpha/2}(n-1)\right\}=1-\alpha.$$

The confidence interval for μ with confidence coefficient $1-\alpha$ is

$$\left(\overline{X}-\frac{S}{\sqrt{n}}t_{\alpha/2}(n-1),\overline{X}+\frac{S}{\sqrt{n}}t_{\alpha/2}(n-1)\right).$$

Example7.4.1

Consider the following sample of fat content (in percentage) of $n=10$ randomly selected hot dogs 25.2, 21.3, 22.8, 17.0, 29.8, 21.0, 25.5, 16.0, 20.9, 19.5. Assuming that these were selected from a normal population distribution, a 95% confidence interval for the population mean fat content is

$$\overline{x}\pm t_{0.025}(9)\frac{s}{\sqrt{n}}=21.9\pm2.262\frac{4.134}{\sqrt{10}}=(18.94,24.86)$$

Suppose, however, you are going to eat a single hot dog of this type and want a prediction for the resulting fat content. A point prediction, analogous to a point estimate, is just $\bar{x}=21.9$. This prediction unfortunately gives no information about reliability or precision.

2. Confidence Interval for σ^2

According to practical problems, here we only discuss the situation of μ is unknown.

We have learned that estimates of population means can be made from sample means, and confidence intervals can be constructed to better describe those estimates. Similarly, we can estimate a population standard deviation from a sample standard deviation, and when the original population is normally distributed, we can construct confidence intervals from the sample variance to describe population variance as well.

The point estimation of σ^2 is S^2, then

$$\chi^2=\frac{(n-1)S^2}{\sigma^2}\sim\chi^2(n-1).$$

From the definition of upper α quantile of χ^2 distribution, then

$$P\left\{\chi^2_{1-\alpha/2}(n-1)<\frac{(n-1)S^2}{\sigma^2}<\chi^2_{\alpha/2}(n-1)\right\}=1-\alpha.$$

That is

$$P\left\{\frac{(n-1)S^2}{\chi^2_{\alpha/2}(n-1)}<\sigma^2<\frac{(n-1)S^2}{\chi^2_{1-\alpha/2}(n-1)}\right\}=1-\alpha.$$

Then we get the confidence interval for σ^2 with confidence coefficient $1-\alpha$ is

$$\left(\frac{(n-1)S^2}{\chi^2_{\alpha/2}(n-1)},\frac{(n-1)S^2}{\chi^2_{1-\alpha/2}(n-1)}\right).$$

At the same time, we get the confidence interval for σ with confidence coefficient $1-\alpha$ is

$$\left(\frac{\sqrt{n-1}S}{\sqrt{\chi^2_{\alpha/2}(n-1)}},\frac{\sqrt{n-1}S}{\sqrt{\chi^2_{1-\alpha/2}(n-1)}}\right).$$

Example7. 4. 2

Suppose the strength of the material has a normal distribution $N(\mu,\sigma^2)$

with parameters μ and σ^2 whose values are unknown, we get 20 units to test the strength, the sample variance of 20 units is $S^2=0.0912$. Find confidence interval for σ^2 with confidence coefficient 0.90.

Solution:

Because $\alpha=0.1$, from table we know $\chi^2_{0.05}(19)=30.144$, $\chi^2_{0.95}(19)=10.117$, the confidence interval for σ^2 with confidence coefficient $1-\alpha$ is

$$\left(\frac{(n-1)S^2}{\chi^2_{\alpha/2}(n-1)}, \frac{(n-1)S^2}{\chi^2_{1-\alpha/2}(n-1)}\right).$$

So we get confidence interval for σ^2 is

$$\left(\frac{19\times 0.0912}{30.144}, \frac{19\times 0.0912}{10.117}\right).$$

That is

$$(0.0575, 0.1713).$$

7.4.2 Interval Estimation of Two Normal Population Parameters

In practice we often encounter the following problems: the quality indicator of a known product has a normal distribution, however, the differences of raw materials, equipment conditions, the operators or factors such as process, may cause changes in population mean and population variance, which we need to know how much these changes are. Then it is important to consider the estimation problems of difference of means and ratio of variances in two normal populations.

Suppose that $X_1, \cdots, X_{n_1}$ and $Y_1, \cdots, Y_{n_2}$ form two random and independent samples from normal populations $N(\mu_1, \sigma_1^2)$ and $N(\mu_2, \sigma_2^2)$ with sample means $\overline{X}=\frac{1}{n_1}\sum_{i=1}^{n_1}X_i$, $\overline{Y}=\frac{1}{n_2}\sum_{i=1}^{n_2}Y_i$ and sample variances $S_1^2=\frac{1}{n_1-1}\sum_{i=1}^{n_1}(X_i-\overline{X})^2$, $S_2^2=\frac{1}{n_2-1}\sum_{i=1}^{n_2}(Y_i-\overline{Y})^2$, respectively. The confidence coefficient $1-\alpha$ is known.

1. the Confidence Interval for $\mu_1-\mu_2$ of Two Populations

(1) When σ_1^2, σ_2^2 are known, because $\overline{X}, \overline{Y}$ are unbiased estimator of μ_1, μ_2

respectively, then

$$U=\frac{\overline{X}-\overline{Y}-(\mu_1-\mu_2)}{\sqrt{\frac{\sigma_1^2}{n_1}+\frac{\sigma_2^2}{n_2}}}\sim N(0,1).$$

The confidence interval for $\mu_1-\mu_2$ with confidence coefficient $1-\alpha$ is

$$\left(\overline{X}-\overline{Y}\pm z_{\alpha/2}\sqrt{\frac{\sigma_1^2}{n_1}+\frac{\sigma_2^2}{n_2}}\right).$$

(2) When σ_1^2,σ_2^2 are unknown, as long as n_1,n_2 are large enough ($n_1>50, n_2>50$), then the confidence interval for $\mu_1-\mu_2$ with confidence coefficient $1-\alpha$ is

$$\left(\overline{X}-\overline{Y}\pm z_{\alpha/2}\sqrt{\frac{S_1^2}{n_1}+\frac{S_2^2}{n_2}}\right).$$

(3) $\sigma_1^2=\sigma_2^2=\sigma^2$, but σ^2 is unknown, from Theorem 6. 4. 3 in Chapter 6 we know

$$t=\frac{\overline{X}-\overline{Y}-(\mu_1-\mu_2)}{S_w\sqrt{\frac{\sigma_1^2}{n_1}+\frac{\sigma_2^2}{n_2}}}\sim t(n_1+n_1-2),$$

where

$$S_w^2=\frac{(n_1-1)S_1^2+(n_2-1)S_2^2}{n_1+n_2-2},\quad S_w=\sqrt{S_w^2}.$$

Then we get the confidence interval for $\mu_1-\mu_2$ with confidence coefficient $1-\alpha$ is

$$\left(\overline{X}-\overline{Y}\pm t_{\alpha/2}(n_1+n_2-2)S_w\sqrt{\frac{1}{n_1}+\frac{1}{n_2}}\right).$$

Example7. 4. 3

There are two kinds of sleeping pills, X and Y indicate the extended sleeping hours after taking the pills respectively, we randomly select 10 patients who take the first kind and 10 patients who take the second kind, and get $\overline{X}=2.33$, $S_1^2=1.9$, $\overline{Y}=1.75$, $S_2^2=2.9$, suppose $X\sim N(\mu_1,\sigma^2)$, $Y\sim N(\mu_2,\sigma^2)$, find the confidence interval for $\mu_1-\mu_2$ with confidence coefficient 0.95.

Solution:

$\alpha=0.05, t_{0.025}(18)=2.101$, from $\left(\overline{X}-\overline{Y}\pm t_{\alpha/2}(n_1+n_2-2)S_w\sqrt{\frac{1}{n_1}+\frac{1}{n_2}}\right)$

we get the confidence interval for $\mu_1-\mu_2$ with confidence coefficient 0.95 is

$$\left(\overline{X}-\overline{Y}\pm t_{\alpha/2}(n_1+n_2-2)S_w\sqrt{\frac{1}{n_1}+\frac{1}{n_2}}\right)$$

$$=\left(2.33-1.75\pm 2.101\times\sqrt{\frac{9\times 1.9+9\times 2.9}{18}}\times\sqrt{\frac{1}{10}+\frac{1}{10}}\right)$$

$$=(1.58\pm 1.46)$$

$$=(0.12, 3.04).$$

Because the confidence interval does not contain zero, then we think the first kind of pill is better than the second one in practice.

2. the Confidence Interval for σ_1^2/σ_2^2 of Two Populations

Here we only discuss the situation of μ_1, μ_2 are unknown, because

$$\frac{(n_1-1)S_1^2}{\sigma_1^2}\sim\chi^2(n_1-1),\quad \frac{(n_2-1)S_2^2}{\sigma_2^2}\sim\chi^2(n_2-1).$$

S_1^2 and S_2^2 are independent, from definition of F distribution,

$$F=\frac{S_1^2/S_2^2}{\sigma_1^2/\sigma_2^2}\sim F(n_1-1, n_2-1).$$

And distribution $F(n_1-1, n_2-1)$ does not depend on any unknown parameters, then

$$P\left\{F_{1-\alpha/2}(n_1-1, n_2-1)<\frac{S_1^2/S_2^2}{\sigma_1^2/\sigma_2^2}<F_{\alpha/2}(n_1-1, n_2-1)\right\}=1-\alpha.$$

That is

$$P\left\{\frac{S_1^2}{S_2^2}\frac{1}{F_{\alpha/2}(n_1-1, n_2-1)}<\frac{\sigma_1^2}{\sigma_2^2}<\frac{S_1^2}{S_2^2}\frac{1}{F_{1-\alpha/2}(n_1-1, n_2-1)}\right\}=1-\alpha,$$

then we get the confidence interval for σ_1^2/σ_2^2 with confidence coefficient $1-\alpha$ is

$$\left(\frac{S_1^2}{S_2^2}\frac{1}{F_{\alpha/2}(n_1-1, n_2-1)}, \frac{S_1^2}{S_2^2}\frac{1}{F_{1-\alpha/2}(n_1-1, n_2-1)}\right).$$

Example 7.4.4

Suppose two group A, B of wire resistance have different normal distributions $X \sim N(\mu_1, \sigma_1^2)$ and $Y \sim N(\mu_2, \sigma_2^2)$ with parameters unknown respectively. Now randomly selecting 6 wires from each group and get resistance as follows (unit: Ω):

Group A: 0.142 0.145 0.143 0.139 0.137 0.141

Group B: 0.140 0.142 0.141 0.138 0.142 0.136

Suppose that the two samples are independent, find the confidence interval for σ_1^2/σ_2^2 with confidence coefficient 0.95.

Solution:

Because μ_1 and μ_2 are unknown, then

$$s_1^2 = 8.167 \times 10^{-6}, \quad s_2^2 = 5.767 \times 10^{-6}, \quad \frac{s_1^2}{s_2^2} = 1.416.$$

And from table we know

$$F_{\alpha/2}(n_1 - 1, n_2 - 1) = F_{0.025}(5,5) = 7.15,$$

$$F_{1-\alpha/2}(n_1 - 1, n_2 - 1) = F_{0.975}(5,5) = 7.15 = \frac{1}{F_{0.025}(5,5)} = 0.140,$$

so the confidence interval for σ_1^2/σ_2^2 with confidence coefficient 0.95 is

$$\left(1.416 \times \frac{1}{7.15}, 1.416 \times \frac{1}{0.14}\right) = (0.198, 10.114).$$

Because the confidence interval for σ_1^2/σ_2^2 contains 1, then we think there is no significant difference between σ_1^2 and σ_2^2.

7.5 One-Sided Confidence Interval

The confidence intervals discussed thus far give both a lower confidence bound and an upper confidence bound for the parameter being estimated. In some circumstances, an investigator will want only one of these two types of bounds. For example, a psychologist may wish to calculate a 95% upper confidence bound for true average reaction time to a particular stimulus, or a reliability engineer may want only a lower confidence bound for true average lifetime of components of a certain type. Because the cumulative area under the curve to the left of 1.645 is 0.95

$$P\left(\frac{\overline{X}-\mu}{S/\sqrt{n}}<1.645\right)\approx 0.95.$$

Manipulating the inequality inside the parentheses to isolate μ on one side and replacing rv's by calculated values gives the inequality $\mu>\overline{x}-1.645s/\sqrt{n}$; the expression on the right is the desired lower confidence bound. Starting with $P(-1.645<z)\approx 0.95$ and manipulating the inequality results in the upper confidence bound. A similar argument gives a one-sided bound associated with any other confidence level.

Definition 7.5.1

As a general case, suppose that $X_1, X_2, \cdots, X_n$ form a random sample from a distribution $F(x;\theta)$ that involves a parameter θ whose value is unknown. Suppose also that the statistic $\underline{\theta}=\underline{\theta}(X_1, X_2, \cdots, X_n)$ can be found such that no matter what the true value of θ may be,

$$P\{\theta>\underline{\theta}\}=1-\alpha,$$

where α is a fixed probability$(0<\alpha<1)$. Then it is said that $(\underline{\theta}, +\infty)$ is a one-sided **confidence interval** for θ with **confidence coefficient** $1-\alpha$, $\underline{\theta}=\underline{\theta}(X_1, X_2, \cdots, X_n)$ is a **one-sided lower confidence bound** for θ with **confidence coefficient** $1-\alpha$.

For the same ,if the statistic $\theta=\overline{\theta}(X_1, X_2, \cdots X_n)$ meet the condition

$$P\{\theta<\overline{\theta}\}=1-\alpha.$$

Then it is said that $(-\infty, \overline{\theta})$ is a one-sided **confidence interval** for θ with **confidence coefficient** $1-\alpha$, $\overline{\theta}=\overline{\theta}(X_1, X_2, \cdots, X_n)$ is a **one-sided upper confidence bound** for θ with **confidence coefficient** $1-\alpha$.

Proposition: A large-sample upper confidence bound for μ is

$$\mu<\overline{x}+z_\alpha\cdot\frac{s}{\sqrt{n}}$$

and **a large-sample lower confidence bound for** μ is

$$\mu>\overline{x}-z_\alpha\cdot\frac{s}{\sqrt{n}}.$$

Example 7.5.1

The slant sheer test is the most widely accepted procedure for assessing the quality of a bond between a repair material and its concrete substrate. The article reported that in one particular investigation, a sample of 48 sheer strength observations gave a sample mean strength of 17.17 N/mm^2 and a sample standard deviation of 3.28 bound . A lower confidence for true average sheer strength μ with confidence level 95% is

$$17.17-(1.645)\frac{3.28}{\sqrt{48}}=17.17-0.78=16.39.$$

That is, with a confidence level of 95%, the value of μ lies in the interval $(16.39, \infty)$.

Example 7.5.2

Randomly selecting 5 bulbs from a group to test the lifetime, the lifetimes(hours) are as follows:

1 050　1 100　1 120　1 250　1 280

Suppose that the lifetime has a normal distribution. Find the one-sided lower confidence bound for μ with confidence coefficient 0.95, where μ is the mean of population.

Solution.

Because $1-\alpha=0.95, n=5, t_\alpha(n-1)=t_{0.05}(4)=2.131\,8, \bar{x}=1\,160, s^2=9\,950$.

Then we get the one-sided lower confidence bound for μ with confidence coefficient 0.95 is

$$\underline{\mu}=\bar{x}-\frac{s}{\sqrt{n}}t_\alpha(n-1)=1\,065.$$

Example 7.5.3

We use balance to weigh a object five times and get the results: 5.52, 5.48, 5.64, 5.51, 5.43. Suppose the results have normal distribution $N(\mu, \sigma^2)$ with parameters μ and σ whose values are unknown, find the one-sided upper confidence interval for σ with confidence coefficient 0.90.

Solution.

Because

$$\chi^2=\frac{(n-1)S^2}{\sigma^2}\sim\chi^2(n-1),$$

for a given α,

$$P\left\{\frac{(n-1)S^2}{\sigma^2}>\chi^2_{1-\alpha}(n-1)\right\}=1-\alpha.$$

That is

$$P\left\{\sigma<\frac{\sqrt{n-1}S}{\sqrt{\chi^2_{1-\alpha}(n-1)}}\right\}=1-\alpha.$$

We can get the one-sided upper confidence interval for σ with confidence coefficient $1-\alpha$ is

$$\bar{\sigma}=\frac{\sqrt{n-1}S}{\sqrt{\chi^2_{1-\alpha}(n-1)}}.$$

Now $n=5$, $S=0.0777$, $\alpha=0.1$. Then from table we get $\chi^2_{0.9}(4)=1.064$, so the one-sided upper confidence interval for σ with confidence coefficient 0.9 is

$$\bar{\sigma}=\frac{\sqrt{4}\times 0.0777}{\sqrt{1.064}}=0.151.$$

Exercises

7.1 Randomly selecting 8 circles, and find the diameters (mm) are

74.001 74.005 74.003 74.001

74.000 73.998 74.006 74.002

(1) Find moment estimation of population mean μ and population variance σ^2.

(2) Find the sample variance s^2.

7.2 Suppose that $X_1, X_2, \cdots, X_n$ form a random sample from population X, $x_1, x_2, \cdots, x_n$ are corresponding observations. Find the moment estimation of unknown parameters below.

(1) $f(x)=\begin{cases}\theta c^{\theta}x^{-(\theta+1)}, & x>c;\\ 0, & \text{otherwise.}\end{cases}$

Where $c>0$, c is a known parameter, $\theta>1$, θ is an unknown parameter.

(2) $f(x)=\begin{cases}\sqrt{\theta}x^{\sqrt{\theta}-1}, & 0\leqslant x\leqslant 1;\\ 0, & \text{otherwise.}\end{cases}$

Where $\theta>0$, θ is an unknown parameter.

(3) $P\{X=x\}=\binom{m}{x}p^x(1-p)^{m-x}$, $x=0,1,2,\cdots,m$, $0<p<1$, p is an unknown parameter.

7.3 Find the M. L. E. of unknown parameters in exercise 7.2 above.

7.4 Suppose that $X_1, X_2, \cdots, X_n$ form a random sample from a Bernoulli distribution for which the parameter θ is unknown, but it is known that θ lies in the open interval $0<\theta<1$. Show that the M. L. E. of θ does not exist if every observed value is 0 or if every observed value is 1.

7.5 Suppose that $X_1, X_2, \cdots, X_n$ form a random sample from a Poisson distribution for which the mean θ is unknown, $(\theta>0)$

(1) Determine the M. L. E. of θ, assuming that at least one of the observed values is different from 0.

(2) Show that the M. L. E. of θ does not exist if every observed value is 0.

7.6 Suppose that $X_1, X_2, \cdots, X_n$ form a random sample from a normal distribution for which the mean μ is known, but the variance σ^2 is unknown. Find the M. L. E. of σ^2.

7.7 Suppose that $X_1, X_2, \cdots, X_n$ form a random sample from an exponential distribution for which the value of the parameter β is unknown $\beta>0$. Find the M. L. E. of β.

7.8 Suppose that $X_1, X_2, \cdots, X_n$ form a random sample from a distribution for which the p. d. f. $f(x;\theta)$ is as follows:

$$f(x;\theta)=\begin{cases}e^{\theta-x}, & x>\theta;\\ 0, & x\leqslant\theta.\end{cases}$$

Also, suppose that the value of θ is unknown $(-\infty<\theta<+\infty)$.

(1) Show that the M. L. E. of θ does not exist.

(2) Determine another version of the p. d. f. of this same distribution for which the M. L. E. of θ will exist, and find this estimator.

7.9 Suppose that $X_1, X_2, \cdots, X_n$ form a random sample from a distribution for which the p. d. f. $f(x;\theta)$ is as follows:

$$f(x;\theta)=\frac{1}{2}e^{-|x-\theta|}, \quad -\infty<x<+\infty.$$

Also, suppose that the value of θ is unknown $-\infty<x<+\infty$. Find the M. L. E. of θ.

7.10 Suppose that $X_1, X_2, \cdots, X_n$ form a random sample from a uniform distribution on the interval $[\theta_1, \theta_2]$, where both θ_1 and θ_2 are unknown($-\infty<\theta_1<\theta_2<+\infty$). Find the M. L. E.' of θ_1 and θ_2.

7.11 Suppose probability distribution of population X is

X	1	2	3
p_k	θ^2	$2\theta(1-\theta)$	$(1-\theta)^2$

θ is unknown $(0<\theta<1)$, because we get the sample values $x_1=1, x_2=2, x_3=1$. Find the moment estimation and M. L. E. of θ.

7.12 Suppose that $X_1, X_2, \cdots, X_n$ form a random sample from a Poisson distribution with parameter λ. Find the Moment estimation and M. L. E. of λ.

7.13 (1) Prove

$$S_w^2=\frac{n_1-1}{n_1+n_2-2}S_1^2+\frac{n_2-1}{n_1+n_2-2}S_2^2=\frac{(n_1-1)S_1^2+(n_2-1)S_2^2}{n_1+n_2-2}$$

is unbiased estimator of σ^2. (Two population have the same variance σ^2)

(2) Suppose that $X_1, X_2, \cdots, X_n$ form a random sample from population$X$, $E(X)=\mu$, $a_1, a_2, \cdots, a_n$ are random constants. Prove $\sum_{i=1}^{n}a_iX_i \Big/ \sum_{i=1}^{n}a_i$ is unbiased estimator of μ ($\sum_{i=1}^{n}a_i \neq 0$).

7.14 Suppose that $X_1, X_2, \cdots, X_n$ form a random sample from a distribution for which the p. d. f. $f(x;\theta)$ is as follows:

$$f(x;\theta)=\begin{cases}\frac{1}{\theta}x^{(1-\theta)/\theta}, & 0<x<1; \\ 0, & \text{otherwise.}\end{cases}$$

Also, suppose that the value of θ is unknown $(0<\theta<\infty)$.

(1) Find the M. L. E. of θ.

(2) Prove M. L. E. $\hat{\theta}$ is unbiased estimator of θ.

7.15 Suppose that X_1, X_2, X_3, X_4 form a random sample from exponential distribution with mean θ, θ is unknown, suppose estimators

$$T_1=\frac{1}{6}(X_1+X_2)+\frac{1}{3}(X_3+X_4),$$

$$T_2=(X_1+2X_2+3X_3+4X_4)/5,$$

$$T_3=(X_1+X_2+X_3+X_4)/4.$$

(1) For T_1, T_2, T_3, which is the unbiased estimator of θ?

(2) For unbiased estimator above, which is more effective?

7.16 Suppose that X_1, X_2, $\cdots$, X_n form a random sample from population X which has a exponential distribution, the p. d. f. of X is

$$f(x)=\begin{cases}\lambda e^{-\lambda x}, & x\geqslant 0;\\ 0, & x<0.\end{cases}$$

(1) Prove $2n\lambda\overline{X}\sim\chi^2(2n)$.

(2) Find the confidence interval for λ with confidence coefficient $1-\alpha$.

(3) Find the one-sided upper confidence interval for λ with confidence coefficient $1-\alpha$.

7.17 Suppose that a random sample of eight observations is taken from a normal distribution for which both the mean μ and the variance σ^2 are unknown; and that the observed values are 3.1, 3.5, 2.6, 3.4, 3.8, 3.0, 2.9 and 2.2. Find the shortest confidence interval for μ with each of the following three confidence coefficients:

(1) 0.90; (2) 0.95; (3) 0.99.

7.18 In the June 1986 issue of Consumer Reports, some data on the calorie content of beef hot dogs is given. Here are the numbers of calories in 20 different hot dog brands:

186, 181, 176, 149, 184, 190, 158, 139, 175, 148,
152, 111, 141, 153, 190, 157, 131, 149, 135, 132

Assume that these numbers are the observed values from a random sample of twenty independent normal random variables with mean μ and variance σ^2, both unknown. Find a 90 percent confidence interval for the mean number of calories μ.

7.19 Suppose that $X_1, X_2, \cdots, X_{10}$ form a random sample from normal population $N(6.5, \sigma^2)$ with unknown parameter σ, the observations are

7.5 2.0 12.1 8.8 9.4 7.3 1.9 2.8 7.0 7.3

Find the confidence interval for σ with confidence coefficient 0.95.

7.20 Randomly selecting 4 wires from group A and 5 wires from group B, and the resistance(Ω) are

A: 0.143 0.142 0.143 0.137

B: 0.140 0.142 0.136 0.138 0.140

Suppose that A has a normal distribution $N(\mu_1, \sigma^2)$ and B has a normal distribution $N(\mu_2, \sigma^2)$, the two samples are independent, μ_1, μ_2 and σ^2 are unknown parameters.

(1) Find the confidence interval for $\mu_1 - \mu_2$ with confidence coefficient 0.95.

(2) Find the one-sided lower confidence interval for $\mu_1 - \mu_2$ with confidence coefficient 0.95.

7.21 Suppose that two conductors A and B test the chemicals 10 times respectively, the sample variances are $s_A^2 = 0.5419$, $s_B^2 = 0.6065$, suppose that σ_A^2 and σ_B^2 are population variance, the two samples are independent and the two populations are normally distributed.

(1) Find the confidence interval for σ_A^2/σ_B^2 with confidence coefficient 0.95.

(2) Find the one-sided upper confidence interval for σ_A^2/σ_B^2 with confidence coefficient 0.95.

Chapter 8 Testing Hypotheses

In statistics, we usually face to the problem like "Does a new medicine work?" "Is one group scoring significantly higher on average than other group?" Frequently the objective of an investigation is not to estimate a parameter but to decide which of two contradictory claims about the parameter is correct. Methods for accomplishing this comprise the part of statistical inference called hypothesis testing.

In this chapter, first we introduce the notation and some common methodology associated with hypothesis testing. Then we use some specific examples to illustrate the way to make a real test of hypothesis. We also demonstrate an equivalence between hypothesis tests and confidence intervals. At last of the chapter we give a brief introduction about p-value.

8.1 Problem of Testing Hypotheses

A statistical hypothesis, or hypothesis, is a claim either about the value of a single population characteristic or about the values of several population characteristics. One example of a hypothesis is the claim $\mu = .75$, where μ is the true average inside diameter of a certain type of PVC pipe. Also, if μ_1 and μ_2 denote the true average breaking strength of two different types of twine, one hypothesis is the assertion that $\mu_1 - \mu_2 = 0$, and the other is the statement

$\mu_1-\mu_2\neq0$. Also one hypothesis can be $\mu_1-\mu_2>5$ and the other one is the statement $\mu_1-\mu_2\leqslant5$. Further kinds of hypothesis can be seen at section 8. 2 and section 8. 3. In any hypothesis-testing problem, there are two contradictory hypotheses under consideration. The objective is to decide, based on same information, which of the two hypotheses is correct.

8. 1. 1 Notations and Concepts

First we will start with an example to see what a real test of hypothesis is:

Example 8. 1. 1

Manly reports measurements of various dimensions of human skulls found in Egypt from various time periods. These data are attributed to Thomson and Randall-Maciver (1905). One time period is approximately 4 000 B. C. We might model the observed breadth measurements (in mm) of the skulls as normal random variables with unknown mean θ and variance 27. Interest might lie in how θ compares to the breadth of a modern-day skull, about 140mm.

In this case, we would write the hypotheses as

$$H_0:\theta\leqslant140,\quad H_1:\theta>140.$$

More realistically, we would assume that both the mean and variance of breadth measurements were unknown. That is, each measurement is a normal random variable with mean μ and variance σ^2. In this case, the parameter would be two-dimensional, for example $\theta=(\mu,\sigma^2)$. Since the hypotheses only concern the first parameter μ. We would then write the hypotheses to be tested as

$$H_0:\mu\leqslant140,\quad H_1:\mu>140.$$

In testing statistical hypotheses, the problem will be formulated so that one of the claims is initially favored. This initially favored claim will not be rejected in favor of the alternative claim unless sample evidence contradicts it and provides strong support for the alternative assertion.

Definition 8. 1. 1

Null hypothesis refers to a general statement or default position of one or more population characteristics. The null hypothesis is generally assumed to be true until evidence indicates otherwise. In statistics, it is often denoted by H_0. The alternative hypothesis, denoted by H_1, is the assertion that is contradictory to H_0.

One way of describing the decisions available to the statistician is that he may accept either H_0 or H_1. However, since there are only two possible decisions, accepting H_0 is equivalent to rejecting H_1, and accepting H_1 is equivalent to rejecting H_0. We shall use all these descriptions in our discussions of testing hypotheses. Some authors use the phrase "do not reject H_0" rather than "accept H_1".

Then we need a tool to help us judge whether a hypothesis is true. Let S denote the sample space of the n-dimensional random vector $X=(X_1,\cdots,X_n)$. In other words, S is the set of all possible outcomes of the samples.

In a problem of this type, the statistician specifies a test procedure by partitioning the sample space S into two subsets. One subset contains the values of X for which she will accept H_0, and the other subset contains the values of X for which she will reject H_0 and therefore accept H_1. The subset for which H_0 will be rejected is called the critical region of the test. In summary, a test procedure is determined by specifying the critical region of the test. The complement of the critical region must then contain all the outcomes for which H_0 will be accepted.

In most hypothesis testing problems, the critical region is defined in terms of a test statistic, $T=r(X)$, a function of the data X. Typically, a test using test statistic T will reject the null hypothesis if T falls in some fixed interval or falls outside of some fixed interval. For example, if the test rejects H_0 when $T\geqslant c$, the critical region is the set of all X such that $r(X)\geqslant c$. Once a test statistic is found, it is simpler to ignore the critical region and just focus on whether or not T is in or out of the fixed interval. All of the tests in the rest of this book will be based on test statistics. Indeed, most of the tests can be written in the form "Reject H_0 if $T \geqslant c$."

Definition 8.1.2

Let C be that subset of the sample space which, in accordance with a prescribed test, leads to rejection of he hypothesis under consideration. Then C is called the critical region of the test.

Example 8.1.2

A workshop are packaging glucose with a packaging machine, here the weight of the glucose is a random variable with a normal distribution. When the machine is normal, the mean of the glucose is 0.5 kg, and the standard deviation is 0.015 kg. One day we want to test if the packaging machine is normally working. We get 9 glucose which is packed by the machine, and here are their weight (kg)

0.497 0.506 0.518 0.524 0.498 0.511 0.520 0.515 0.512.

Suppose μ and σ are the mean and standard deviation of the population X. Practice shows that the variance is stable considering the long-term, so let $\sigma = 0.015$. Then $X \sim N(\mu, 0.015^2)$, here μ is unknown, we have to make a choice whether $\mu = 0.5$ or not, and give the following hypothesis:

$$H_0: \mu = \mu_0 = 0.5, \quad H_1: \mu \neq \mu_0.$$

We need to give a test statistic and make judgment by the law if we should accept H_0. Considering the test statistic, if we decide to accept H_0, that is to say, the machine is normally working.

8.1.2 Types of Error and Significance Level

Since the criterion of testing is according to the samples, it is likely to make wrong decision. When hypothesis H_0 is true, it has become traditional to call an erroneous decision to reject a true null hypothesis H_0 type 1 error, or an error of the first kind. But when hypothesis H_0 is not true, an erroneous decision to accept a false null hypothesis is called a type 2 error, or an error of the second kind.

Table 8.1.1

	Null true	Null false
Accept Null Hypothesis	Correct Decision	Type 2 error
Reject Null Hypothesis	Type 1 error	Correct Decision

Instead of demanding error-free procedures, we must look for procedures for which either type of error is unlikely to occur. That is, a good procedure is one for which the probability of making either type of error is small. The choice of a particular rejection region cutoff value fixes the probability of type 1 and type 2 errors. These error probability are traditional denoted by α and β, respectively. Because H_0 specifies a unique value of the parameters, there is a single value of α. However, there is a different value of β for each value of the parameter consistent with H_1.

According to the upper statement, when we decide the testing criterion, we should decrease the probability of making two errors as far as possible. But when the sample size is fixed, the decreasing of one type of error means the increasing of another type of error. In general, we always choose to control the first kind error and make it not more than α. Usually α is equal 0.1, 0.05, 0.01, 0.005. The test of controlling the first kind of error and not considering the probability of making the second type of error is called significant test.

In other word, the significance level is the criterion used for rejecting the null hypothesis. The significance level is used in hypothesis testing as follows: First, the difference between the results of the experiment and the null hypothesis is determined. Then, assuming the null hypothesis is true, the probability of a difference that large or larger is computed. Finally, this probability is compared to the significance level. If the probability is less than or equal to the significance level, then the null hypothesis is rejected and the outcome is said to be statistically significant. Traditionally, experimenters have used either the 0.05 level (sometimes called the 5% level) or the 0.01

level (1% level), although the choice of levels is largely subjective. The lower the significance level, the more the data must diverge from the null hypothesis to be significant. Therefore, the 0.01 level is more conservative than the 0.05 level.

Here we will use an example to illustrate the problem.

Example 8.1.3 (Consider the test in Example 8.1.2)

We use the test statistic $Z=\dfrac{\overline{X}-\mu_0}{\sigma/\sqrt{n}}$. When we choose to control the first kind error and make it not more than α, that is to say we want:

$$P\{\text{reject } H_0 \text{ when } H_0 \text{ is true}\}\leqslant\alpha$$

We need to find k and let

$$P\{\text{reject } H_0 \text{ when } H_0 \text{ is true}\}=P_{\mu_0}\left\{\left|\frac{\overline{X}-\mu_0}{\sigma/\sqrt{n}}\right|\geqslant k\right\}=\alpha.$$

For $Z=\dfrac{\overline{X}-\mu_0}{\sigma/\sqrt{n}}\sim N(0,1)$. we know $k=z_{\alpha/2}$. α can be seen in Figure 8.1.1.

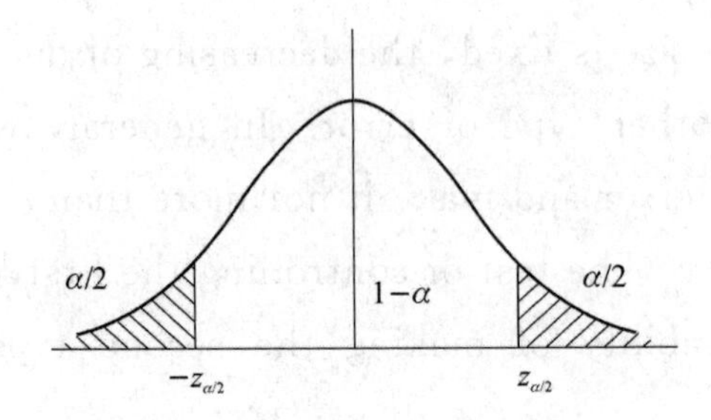

Figure 8.1.1

Example 8.1.4

The target thickness for silicon wafers used in a certain type of integrated circuit is 245 μm. A sample of 50 wafers is obtained and the thickness of each one is determined, resulting in a sample mean thickness of 246.18 μm and a sample standard deviation of 3.6 μm. Does this data suggest that the true average wafer thickness is something than the target value?

Solution:

$$H_0: \mu=245, \quad H_1: \mu\neq 245.$$

Statistic value:

$$z = \frac{\overline{X} - 245}{s/\sqrt{n}}.$$

Calculation of z, we get $z = 2.32$, after the determination of P-value: because the test is two-tailed,

$$P\text{-value} = 2(1 - \phi(2.32)) = 0.0204.$$

Since $\alpha = 0.05$ is larger than the P-value, H_0 would be rejected by anyone carrying out the test at level 0.05. However, at level 0.01, H_0 would not be rejected because 0.01 is smallest level 0.038 4 at which H_0 can be rejected.

8.2 the Testing of Hypotheses of the Mean of the Normal Distribution

First we will determine the following steps in testing hypotheses. Establish hypotheses: state the null and alternative hypotheses. Determine the appropriate statistical test and sampling distribution. Specify the Type I error rate . State the decision rule. Gather sample data. Calculate the value of the test statistic. State the statistical conclusion. Make a managerial decision. Then we may say several specific kinds of test in the following section.

8.2.1 Testing Hypotheses about the Mean of a Normal Distribution with Known Variance

Although the assumption that the value of σ is known is rarely met in practice, this case provides a good starting point because of the ease with which general procedures and their properties can be developed.

Supposing a random sample $X_1, X_2, \cdots, X_n$ is drawn from a normal distribution $N(\mu, \sigma^2)$ with unknown μ and known σ^2.

When testing hypotheses about the mean of a normal distribution, it is traditional to rewrite this test in terms of the statistic

$$Z=\frac{\overline{X}-\mu_0}{\sigma/\sqrt{n}}.$$

Considering when H_0 is true, Z follows the normal distribution with a mean of 0 and a variance of 1, knowing this, we can build a rejection region of Z. That is to say if the null hypothesis is $H_0: \mu=\mu_0$, we will reject it when $|Z| \geqslant \Phi^{-1}(1-\alpha_0/2)$. The method of testing above is called Z test.

Table 8.2.1 gives a summary of Z test.

Table 8.2.1

Null hypothesis: $H_0: \mu=\mu_0$ Test statistic value: $z=\frac{\overline{x}-\mu_0}{\sigma/\sqrt{n}}$	
Alternative Hypothesis	Rejection Region for Level α
$H_1: \mu>\mu_0$	$z \geqslant z_\alpha$
$H_1: \mu<\mu_0$	$z \leqslant z_\alpha$
$H_1: \mu \neq \mu_0$	$\|z\| \geqslant z_{\alpha/2}$

Example 8.2.1 Testing the milk with water

A company bought milk from a manufacturer. The company doubted that the manufacturer made a profit in the milk with water. Through measuring the freezing point of the milk, they can test whether the milk is with water. The freezing point temperature of milk is approximately normal distribution with mean $\mu_0=-0.545$℃ and standard variance $\sigma=0.008$℃. The freezing point of milk with water would increase to the freezing point of water (0℃). Now they test the freezing point of the five bitch of milk, and the mean of this sample is $\overline{x}=-0.535$℃. We want to know whether the manufacturer mixed water with milk. $\alpha=0.05$.

Suppose that we wish to test the hypotheses:

$$H_0: \mu \leqslant \mu_0, \quad H_1: \mu > \mu_0,$$

this is the right-side testing. H_0 means that the milk is not mixed with water. And H_1 means that the milk is mixed with water. According to the Z test, the reject region of right-side hypotheses is

$$z = \frac{\overline{x} - \mu_0}{\sigma/\sqrt{n}} \geqslant z_{0.05} = 1.645.$$

Now $z = \dfrac{-0.535 - (-0.545)}{0.008/\sqrt{5}} = 2.7951 > 1.645$. The value of z is in the reject region. Therefore we can consider that reject H_0 under the level of significance $\alpha = 0.05$. That is, the manufacturer mixed the water with milk.

8.2.2 Testing Hypotheses about the Mean of a Normal Distribution with unknown Variance

A random sample $X_1, X_2, \cdots, X_n$ is drawn from a normal distribution $N(\mu, \sigma^2)$ with unknown μ and unknown σ^2. Since the variance of this distribution is unknown, we can't use the Z statistics. Notice that S^2 are the unbiased estimator of σ^2, we substitute S for σ and use

$$t = \frac{\overline{X} - \mu_0}{S/\sqrt{n}}$$

to be the testing statistics.

If the hypothesis testing is

$$H_0: \mu = \mu_0 \quad H_1: \mu \neq \mu_0.$$

When the observed value $|t| = \left|\dfrac{\overline{x} - \mu_0}{s/\sqrt{n}}\right|$ is big enough, H_0 is needed to be rejected. The form of reject region is

$$|t| = \frac{\overline{x} - \mu_0}{s/\sqrt{n}} \geqslant k.$$

The method of testing above is called t test.

Table 8.2.2

Null hypothesis: $H_0: \mu=\mu_0$ Test statistic value: $t=\frac{\bar{x}-\mu_0}{s/\sqrt{n}}$	
Alternative Hypothesis	Rejection Region for Level α
$H_1: \mu>\mu_0$	$t \geqslant t_\alpha(n-1)$
$H_1: \mu<\mu_0$	$t \leqslant -t_\alpha(n-1)$
$H_1: \mu \neq \mu_0$	$\lvert t \rvert \geqslant t_{\alpha/2}(n-1)$

Example 8.2.2 The testing of the life of one component.

The life X of the component is normal distribution $N(\mu, \sigma^2)$ with unknown μ and unknown σ^2. Now the life of 16 components is as follows:

159 280 101 212 224 379 179 264

222 362 168 250 149 260 485 170

Whether the mean of the life of the component is larger than 225h?

Suppose that we wish that to test the hypotheses

$$H_0: \mu \leqslant \mu_0 = 225,$$
$$H_1: \mu > 225,$$

with level of the significance $\alpha = 0.05$. According to the t test, the reject region of right-side hypotheses is

$$t=\frac{\bar{x}-\mu_0}{s/\sqrt{n}} \geqslant t_\alpha(n-1).$$

Now we have $n=16$, $t_{0.05}(15)=1.7531$ and $\bar{x}=241.5$, $s=98.7259$. That is,

$$t=\frac{\bar{x}-\mu_0}{s/\sqrt{n}}=0.6685<1.7531.$$

The value of t is not in reject region. So we accept H_0 and consider that the mean of the life of the component is not larger than 225h.

8.2.3 The Two-Sample U-Test

Consider the easiest problem in the two-sample test: When the random samples are available from two normal distributions with common known variance. we shall assume that the variables $X_1, \cdots, X_m$ form a random sample of m observations from a normal distribution for which the mean μ_1 is unknown and the variance σ_1^2 is unknown; and that the variables $Y_1, \cdots, Y_n$ form an independent random sample of n observations from another normal distribution for which the mean μ_2 is known and the variance σ_2^2 is unknown. we want to test the value of $\mu_1 - \mu_2$.

Now the hypothesis testing is:

$$H_0: \mu_1 - \mu_2 = \delta, \quad H_1: \mu_1 - \mu_2 \neq \delta.$$

(δ is a known constant and the significance level of testing is α.)

We use the following U-statistics to be our testing statistics:

$$U = \frac{\overline{X} - \overline{Y} - \delta}{\sqrt{\frac{\sigma_1^2}{m} + \frac{\sigma_2^2}{n}}}.$$

When H_0 is true, $U \sim N(0,1)$, we know that the form of reject region $|u| > \mu_{\alpha/2}$. Similarly, other hypothesis and their reject region can be seen as:

Table 8.2.3

H_0	H_1	Reject region
$H_0: \mu_1 - \mu_2 = \delta$	$H_1: \mu_1 - \mu_2 \neq \delta$	$\lvert u \rvert \geqslant \mu_{\alpha/2}$
$H_0: \mu_1 - \mu_2 = \delta$	$H_1: \mu_1 - \mu_2 > \delta$	$u \geqslant \mu_\alpha$
$H_0: \mu_1 - \mu_2 = \delta$	$H_1: \mu_1 - \mu_2 < \delta$	$u \leqslant -\mu_\alpha$

8.2.4 The Two-Sample t-Test

We shall now consider a problem in which random samples are available from two normal distributions with common unknown variance, and it is desired to determine which distribution has the larger mean. Specifically, we shall assume that the variables $X_1, \cdots, X_m$ form a random sample of m observations from a normal distribution for which both the mean μ_1 and the variance σ^2 are unknown; and that the variables $Y_1, \cdots, Y_n$ form an independent random sample of n observations from another normal distribution for which both the mean μ_2 and the variance σ^2 are unknown. We shall assume that the variance σ^2 is the same for both distributions, even though the value of σ^2 is unknown. If this final assumption seems unwarranted, the two-sample t test would not be appropriate. A different test procedure is discussed later in this section for the case in which the two populations might have different variances. In the following part we discuss some procedures for comparing the variances of two normal distributions, which includes testing the hypothesis that the variances are the same.

Now the hypothesis testing is

$$H_0: \mu_1 = \mu_2, \quad H_1: \mu_1 \neq \mu_2.$$

Let the significance level of testing is α.

We use the following t-statistics to be our testing statistics:

$$t = \frac{\overline{X} - \overline{Y}}{S_w \sqrt{\frac{1}{m} + \frac{1}{n}}},$$

where

$$S_w^2 = \frac{(m-1)S_x^2 + (n-1)S_y^2}{m+n-2}, \quad S_w = \sqrt{S_w^2}.$$

When H_0 is true, we know that the form of reject region is

$$\left| \frac{|\overline{x} - \overline{y}|}{S_w \sqrt{\frac{1}{m} + \frac{1}{n}}} \right| \geqslant k.$$

When we set the significance level of testing is α, the reject region is

$$|t| = \frac{|\overline{x} - \overline{y}|}{S_w \sqrt{\frac{1}{m} + \frac{1}{n}}} \geqslant t_{\alpha/2}(m+n-2).$$

Table 8.2.4

Null hypothesis: $H_0: \mu = \mu_0$

Test statistic value: $t = \frac{|\overline{x} - \overline{y}|}{S_w \sqrt{\frac{1}{m} + \frac{1}{n}}}$

Alternative Hypothesis	Rejection Region for Level α
$H_1: \mu > \mu_0$	$t \geqslant t_\alpha(m+n-2)$
$H_1: \mu < \mu_0$	$t \leqslant -t_\alpha(m+n-2)$
$H_1: \mu \neq \mu_0$	$\lvert t\rvert \geqslant t_{\alpha/2}(m+n-2)$

Example 8.2.3 Roman Pottery in Britain.

Tubb, Parker, and Nickless (1980) describe a study of samples of pottery from the Roman era found in various locations in Great Britain. One measurement made on each sample of pottery was the percentage of the sample that was aluminum oxide. Suppose that we are interested in comparing the aluminum oxide percentages at two different locations. There were $m=14$ samples analyzed from Llanederyn with sample average of $\overline{X}_m=12.56$ and $S_X^2=24.65$. Another $n=5$ samples came from Ashley Rails with $\overline{Y}_n=17.32$ and $S_Y^2=11.01$.

Suppose that we model the data as normal random variables with two different means μ_1 and μ_2 but common variance σ^2. We want to test the null hypothesis $H_0: \mu_1 > \mu_2$ against the alternative hypothesis $H_1: \mu_1 < \mu_2$. The observed value of t defined by the former part is -6.302. From the table of the t distribution in this book with $m+n-2=17$ degrees of freedom, we find that $t_{0.995}(17)=2.898$, and $t<-2.898$. So, we would reject H_0 at any level $\alpha_0 \geqslant 0.005$.

8.2.5 The test based on paired data

Two data sets are "paired" when the following one-to-one relationship exists between values in the two data sets.

(1) Each data set has the same number of data points.

(2) Each data point in one data set is related to one, and only one, data point in the other data set.

An example of paired data would be a before-after drug test. The researcher might record the blood pressure of each subject in the study, before and after a drug is administered. These measurements would be paired data, since each "before" measure is related only to the "after" measure from the same subject.

There is n pair mutually independent observed result: (X_1, Y_1), $(X_2, Y_2), \cdots, (X_n, Y_n)$. Let $D_1 = X_1 - Y_1, D_2 = X_2 - Y_2, \cdots, D_n = X_n - Y_n$.

And since that $D_1, D_2, \cdots, D_n$ are caused by the same reason, we can consider the come from the same distribution. Now suppose that $D_i \sim N(\mu_D, \sigma_D^2), i=1,2,\cdots,n$. In other words, $D_1, D_2, \cdots, D_n$ is a sample from normal distribution $N(\mu_D, \sigma_D^2)$, where μ_D, σ_D^2 are unknown. We have the hypothesis test based on this sample: (1) $H_0: \mu_D = 0, H_1: \mu_D \neq 0$; (2) $H_0: \mu_D \leqslant 0, H_1: \mu_D > 0$; (3) $H_0: \mu_D \geqslant 0, H_1: \mu_D < 0$.

Notice that the sample mean and sample variance of sample $D_1, D_2, \cdots, D_n$ are $\overline{d}, s_D^2$ independently. From the definition of t test, the reject region of the former three test (the significance is α) are $|t| = \left|\frac{\overline{d}}{s_D/\sqrt{n}}\right| \geqslant t_{\alpha/2}(n-1)$, $t = \frac{\overline{d}}{s_D/\sqrt{n}} \geqslant t_\alpha(n-1)$, and $t = \frac{\overline{d}}{s_D/\sqrt{n}} \leqslant -t_\alpha(n-1)$.

Example 8.2.4 Reaction time of red light or green light.

The experiment is want to compare the human reaction time of red light or green light (the unit of time is second). The experiment started the chronoscope with red light or green light. If he participants saw the red light or green light, click the button and then the chronoscope stopped. The time interval between the two records of the chronoscope is the reaction time. The result is as follows.

Table 8.2.5

Red light	0.30	0.23	0.41	0.53	0.24	0.36	0.38	0.51
Green light	0.43	0.32	0.58	0.46	0.27	0.41	0.38	0.61
Difference	−0.13	−0.09	−0.17	0.07	−0.03	−0.05	0.00	−0.10

Note that $D_i = X_i - Y_i (i=1,2,\cdots,8)$. And we assume that the sample is normal distribution $N(\mu_D, \sigma_D^2)$ with unknown μ_D and σ_D^2 (the significance level α is 0.05). We need to test hypothesis

$$H_0: \mu_D \geqslant 0, \quad H_1: \mu_D < 0.$$

Now we have $n=8, \overline{x}_d = -0.0625, s_d = 0.0765$,

$$\frac{\bar{x}_d}{s_d/\sqrt{n}}=-2.311<-t_{0.05}(7)=-1.8946.$$

Therefore we should reject H_0. That is, we consider the reaction time of red light is less than the reaction time of green light.

8.3 Testing Hypotheses about Variance of Normal Distribution

8.3.1 Test of variancefor a single population

A random sample $X_1, X_2, \cdots, X_n$ are drawn from a normal distribution $N(\mu, \sigma^2)$ with unknown parameters μ and σ^2. We need to test hypothesis (the level of significance is α)

$$H_0: \sigma^2=\sigma_0^2, \quad H_1: \sigma^2 \neq \sigma_0^2.$$

Here σ_0^2 are known constant.

S^2 is unbiased estimator of σ^2. When H_0 is true, the ratio between the observed value s^2 and σ_0^2 should be fluctuate around 1, not much larger or smaller than 1. We know that $\frac{(n-1)S^2}{\sigma_0^2} \sim \chi^2(n-1)$. Then we get $\chi^2=\frac{(n-1)S^2}{\sigma_0^2}$ to be the test statistics. The form of reject region of the testing problem above is as follows:

$$\frac{(n-1)S^2}{\sigma_0^2} \leqslant k_1, \quad \frac{(n-1)S^2}{\sigma_0^2} \geqslant k_2.$$

Here the values of k_1, k_2 is determined by the following equation:

$$P(H_0 \text{ is true and reject } H_0)=P\left(\frac{(n-1)S^2}{\sigma_0^2} \leqslant k_1 \cup \frac{(n-1)S^2}{\sigma_0^2} \geqslant k_2\right)=\alpha.$$

In order to calculate conveniently, we have

$$P\left(\frac{(n-1)S^2}{\sigma_0^2} \leqslant k_1\right)=\frac{\alpha}{2}, \quad P\left(\frac{(n-1)S^2}{\sigma_0^2} \geqslant k_2\right)=\frac{\alpha}{2}.$$

Then $k_1=\chi^2_{1-\alpha/2}(n-1), k_2=\chi^2_{\alpha/2}(n-1)$. Therefore the reject region is

$$\frac{(n-1)S^2}{\sigma_0^2}\leqslant\chi^2_{1-\alpha/2}(n-1) \text{ or } \frac{(n-1)S^2}{\sigma_0^2}\geqslant\chi^2_{\alpha/2}(n-1).$$

The following problem is one-side test (the level of significance is α)

$$H_0:\sigma^2\leqslant\sigma_0^2,\quad H_1:\sigma^2>\sigma_0^2.$$

When H_0 is true, it can be clarified that $\frac{(n-1)S^2}{\sigma_0^2}\sim\chi^2(n-1)$, we can use it as a test statistic value. Easily we can know, the reject region of the hypothesis $H_0:\sigma^2\geqslant\sigma_0^2$, $H_1:\sigma^2<\sigma_0^2$ is $\sigma^2=\frac{(n-1)s^2}{\sigma_0^2}\leqslant\chi^2_{1-\alpha}(n-1)$. The testing method above is called χ^2 test.

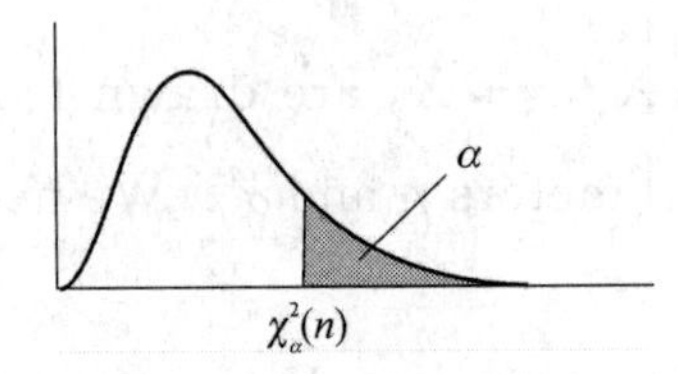

Figure 8. 3. 1

Table 8. 3. 1

Null hypothesis: $H_0:\sigma^2=\sigma_0^2$

Test statistic value: $\frac{(n-1)s^2}{\sigma_0^2}$

Alternative Hypothesis	Rejection Region for Level α
$H_1:\sigma^2\leqslant\sigma_0^2$	$\chi^2<\chi^2_{1-\alpha}(n-1)$
$H_1:\sigma^2\geqslant\sigma_0^2$	$\chi^2>\chi^2_{\alpha}(n-1)$
$H_1:\sigma^2\neq\sigma_0^2$	$\chi^2<\chi^2_{1-\alpha/2}(n-1)$ or $\chi^2>\chi^2_{\alpha/2}(n-1)$

Example 8. 3. 1 The fluctuation of battery time.

A manufacturer produces a kind of battery. For a long time the life time (the unit is hour) is normal distribution with known variance $\sigma^2=5\ 000$. There is a new batch of this kind of battery. The usage of them shows that the fluctuation of the battery time has changed. Now we take 26 battery randomly and work out that the variance of this sample is equal to 9 200. We shall test whether there is a significant change. (The level of significance α is 0.02)

We shall carry out an χ test of the hypothesis

$$H_0: \sigma^2 = 5\,000 \quad H_1: \sigma^2 \neq 5\,000.$$

We have $n=26$, $\chi^2_{\alpha/2}(n-1)=\chi^2_{0.01}(25)=44.314$, $\chi^2_{1-\alpha/2}(n-1)=\chi^2_{0.99}(25)=11.524$. We have known that $\sigma_0^2=5\,000$. Therefore the reject region is

$$\frac{(n-1)s^2}{\sigma_0^2} \geqslant 44.314 \text{ or } \frac{(n-1)s^2}{\sigma_0^2} \leqslant 11.524.$$

According the observed value $s^2 = 9\,200$, we have $\frac{(n-1)s^2}{\sigma_0^2} = 46 > 44.314$. Therefore reject H_0, and consider that the fluctuation of this batch of battery has significant change.

8.3.2 Test of variances for two populations

Suppose that the random variables $X_1, \cdots, X_m$ form a random sample of m observations from a normal distribution for which both the mean μ_1 and the variance σ_1^2 are unknown; and also that the random variables $Y_1, \cdots, Y_n$ form an independent random sample of n observations from another normal distribution for which both the mean μ_2 and the variance σ_2^2 are unknown. Suppose finally that the following hypotheses are to be tested at a specified level of significance α_0 $(0<\alpha_0<1)$:

$$H_0: \sigma_1^2 < \sigma_2^2, \quad H_1: \sigma_1^2 > \sigma_2^2 \tag{8.3.1}$$

Define S_X^2 and S_Y^2 to be the sample variances $S_X^2 = \frac{1}{m-1}\sum_{i=1}^{m}(X_i - \overline{X}_m)^2$ and $S_Y^2 = \frac{1}{n-1}\sum_{i=1}^{n}(Y_i - \overline{Y}_n)^2$, they are estimators of σ_1^2 and σ_2^2, respectively.

It makes intuitive sense that we should reject H_0 if F is large, where $F = \frac{S_X^2}{S_Y^2}$. Such a test is called an F test because the distribution of F is an F distribution when $\sigma_1^2 = \sigma_2^2$.

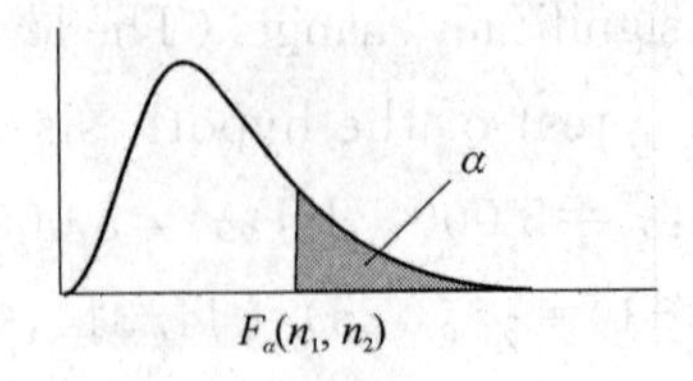

Figure 8. 3. 2

Table 8. 3. 2

Null hypothesis: $H_0: \sigma_1^2=\sigma_2^2$ Test statistic value: $F=S_X^2/S_Y^2$.	
Alternative Hypothesis	Rejection Region for Level α
$H_1: \sigma_1^2>\sigma_2^2$	$F \geqslant F_\alpha(m-1, n-1)$
$H_1: \sigma_1^2<\sigma_2^2$	$F \leqslant F_{1-\alpha}(m-1, n-1)$
$H_1: \sigma_1^2 \neq \sigma_2^2$	$F \geqslant F_{\alpha/2}(m-1, n-1)$ or $F \leqslant F_{1-\alpha/2}(m-1, n-1)$

Example 8. 3. 2

Suppose that six observations $X_1, \cdots, X_6$ are selected at random from a normal distribution for which both the mean μ_1 and the variance σ_1^2 are unknown, and it is found that $S_X^2=5$. Suppose also that 21 observations, $Y_1, \cdots, Y_{21}$, are selected at random from another normal distribution for which both the mean μ_2 and the variance σ_2^2 are unknown; and it is found that $S_Y^2=2$. We shall carry out an F test of the hypotheses.

In this example, $m=6$ and $n=21$, Therefore, when H_0 is true, the statistic F will have an F distribution with 5 and 20 degrees of freedom, the value of F for the given samples is $F=\frac{6}{2}=3$.

It is found from the tables given at the end of this book that the 0. 95 quantile of the F distribution with 5 and 20 degrees of freedom is 2. 71, and the 0. 975 quantile of that distribution is 3. 29. Hence, the tail area corresponding to the value $F=3$ is less than 0. 05 and greater than 0. 025. The hypothesis H_0 that $\sigma_1^2 \leqslant \sigma_2^2$ would therefore be rejected at the level of significance $\alpha_0=0.05$, and

H_0 would be accepted at the level of significance $\alpha_0 = 0.025$.

8.4 Equivalence of Tests and Confidence Sets

There is a strong relationship between tests and confidence sets. We will illustrate the equivalence using some specific cases. Supposing $X_1, \cdots, X_n$ are from a same population and the samples are $x_1, \cdots, x_n$, θ is the parameter which belong to a set Θ.

Then considering (θ_1, θ_2) is a confidence set of θ with a $1-\alpha$ confidence. Here $\theta_1 = \underline{\theta}(X_1, \cdots, X_n)$ and $\theta_2 = \bar{\theta}(X_1, \cdots, X_n)$. For any $\theta \subset \Theta$, we have

$$P(\theta_1 < \theta < \theta_2) \geqslant 1-\alpha. \tag{8.4.1}$$

Considering a test whose significance level is α

$$H_0: \theta = \theta_0, \quad H_1: \theta \neq \theta_0 \tag{8.4.2}$$

From formula (8.4.1) we got

$$P_{\theta_0}(\theta_1 < \theta_0 < \theta_2) \geqslant 1-\alpha.$$

That is to say

$$P_{\theta_0}(\theta_0 \leqslant \theta_1 \cup \theta_0 \geqslant \theta_2) \leqslant \alpha.$$

So the critical region of the test are

$$\theta_0 \leqslant \theta_1 \text{ or } \theta_0 \geqslant \theta_2.$$

It's easy to see when we are testing formula (8.4.1), we first find aconfidence set of θ with a $1-\alpha$ confidence, and then determine whether (θ_1, θ_2) contains θ_0, if so, we will accept the hypothesis H_0. On the other hand, consider a test of θ_0:

$$H_0: \theta = \theta_0, \quad H_1: \theta \neq \theta_0$$

Supposing it'sacceptance region is $\theta_1 < \theta_0 < \theta_2$.

That means $P_{\theta_0}(\theta_1 < \theta_0 < \theta_2) \geqslant 1-\alpha$, for any θ_0, so we get for any $\theta \subset \Theta$, $P_\theta(\theta_1 < \theta < \theta_2) \geqslant 1-\alpha$, So (θ_1, θ_2) is a confidence interval of θ with a $1-\alpha$ confidence level.

Example 8.4.1

Supposing $X \sim N(\mu, 1)$, μ is unknown, $\alpha = 0.05$ and $n = 16$, from a sample we have got $\overline{x}=5.20$ so we got a confidence interval of μ with 95% confidence level is:

$$\left(\overline{x}-\frac{1}{\sqrt{16}}z_{0.025}, \overline{x}+\frac{1}{\sqrt{16}}z_{0.025}\right)=(4.71, 5.69).$$

When talking about the testing problem $H_0: \mu=5.5$, $H_1: \mu\neq 5.5$, we find $5.5\in(4.71, 5.69)$, so we should accept H_0.

8.5 Test of Fit of Population Distribution

All kinds of test method described earlier has been discussed under the condition that the population distribution is known. But in reality, we can't predict the type of population distribution. Then should test the hypothesis of the population distribution based on the samples. This section introduces Pearson χ^2 (K. Pearson) test.

The distribution function $F_n(x)$ of samples $X_i(i=1,2,\cdots,n)$ is known. If choose one distribution function to fit population distribution $F(x)$, there exits some difference between $F(x)$ and $F_n(x)$ no matter what kind of population distribution. Then the question is "the difference is only due to the random error caused by limited number of tests? Or due to the difference between the chosen distribution function $F(x)$ and the known sample distribution $F_n(x)$". For this question, here give the Pearson χ^2 test.

Table 8.5.1 is the frequency of the sample observations from independent trial.

Table 8.5.1

subinterval	Number of frequency	Frequency	probability
$a_0\sim a_1$	n_1	f_1	p_1
$a_1\sim a_2$	n_2	f_2	p_2
$\vdots$	$\vdots$	$\vdots$	$\vdots$
$a_{l-1}\sim a_l$	n_l	f_l	p_l
Total	n	1	1

The null hypothesis is $H_0: F(x)=F_0(x)$, here $F_0(x)$ is one kind of known population distribution. Under the condition that the null hypothesis hold, we will calculate the probability of sample points located in each subinterval

$$p_i=F_0(a_i)-F_0(a_{i-1}),\quad i=1,2,\cdots,l.$$

For testing the null hypothesis H_0, the weighted sum of square of deviation f_i-p_i,

$$Q=\sum_{i=1}^{l} c_i(f_i-p_i)^2.$$

This equation is treated as the measure of deviation between null hypothesis $F_0(x)$ and the sample distribution $F_n(x)$, where c_i is the weight of f_i-p_i. The introducing of weight c_i is necessary, and reasonable, since the deviation between frequency f_i and probability p_i in each subinterval can't be treated equally in the general case. In fact, for the deviation of same absolute value of f_i-p_i, it is not significant when p_i is relatively large, and is very significant when p_i is relatively small. Therefore, the weight c_i is inversely proportional to the probability p_i.

The famous British statisticians Karl Pearson(1951—1936) proved that if $c_i=\frac{n}{p_i}$ and $n\to\infty$, the asymptotic distribution of statistics Q is $\chi^2(l-m-1)$ where l is the number of subinterval and m is the number of unknown parameter of sample observation estimation. The statistics is as follows:

$$\chi^2=\sum_{i=1}^{l}\frac{n(f_i-p_i)^2}{p_i}.$$

We know that $f_i=\frac{n_i}{n}$, $\sum_{i=1}^{l} n_i=n$, $\sum_{i=1}^{l} p_i=1$ For convenience, this equation can be transformed to

$$\begin{aligned}\chi^2&=\sum_{i=1}^{l}\frac{(n_i-np_i)^2}{np_i}\\&=\frac{1}{n}\sum_{i=1}^{l}\frac{n_i^2}{p_i}-2\sum_{i=1}^{l}n_i+n\sum_{i=1}^{l}p_i\\&=\frac{1}{n}\sum_{i=1}^{l}\frac{n_i^2}{p_i}-n.\end{aligned}$$

For the given significance level α,

$$P\{\chi^2 > \chi^2_\alpha(l-m-1)\} = \alpha,$$

the value of $\chi^2_\alpha(l-m-1)$ can be checked in distribution Table.

If the calculated value of statistics χ^2 from the sample observation is larger than $\chi^2_\alpha(l-m-1)$, the null hypothesis H_0 is rejected under significance level α. Otherwise, accept the null hypothesis.

The point which should be paid attention is that the frequency number n_i of the observation located in each subinterval need to be satisfied the condition that the total number is larger than 50 and each frequency number n_i is larger than 5 $(i=1, 2, \cdots, l)$. If the frequency number n_i of some subinterval is too small, we should combine two or more adjacent subinterval to satisfy the frequency number is large enough. Certainly, the number of subinterval l decreases, the freedom of statistics χ^2 will decrease correspondingly.

In addition, when there exits unknown parameter in the null hypothesis distribution function $F_0(x)$, we should estimate these parameter using the method of maximum likelihood estimation.

Example 8.5.1

Here we give an example doing the test of population distribution. The data listed in Table 8.5.2 is the time of rain storm in Shanghai of 63 years between 1875 and 1955.

Table 8.5.2

Rain storm x_i	0	1	2	3	4	5	6	7	8	$\geqslant 9$
Frequency n_i	4	8	14	10	10	4	2	1	1	0

We want to know whether the time of rain storm X's distribution has a Poisson's distribution? $(\alpha=0.05)$

In this example, we want to test $H_0: X \sim p(\lambda)$. The $\lambda > 0$ here is an unknown parameter. First, we should calculate the MLE of λ when H_0 is true:

$$\hat{\lambda}=\bar{x}=\frac{1}{63}(0\times4+1\times8+\cdots+7\times1+8\times1+0)$$

$$=\frac{180}{63}=2.857\,1.$$

Now we use the χ^2 fit test to test

$$H_0: X\sim p(2.857\,1).$$

The probability function is:

$$p(X=i)=\frac{(2.857\,1)^i}{i\,!}e^{-2.856\,1},(i=0,1,2,\cdots)$$

to calculate the observed value of χ^2, we calculate:

Table 8.5.3

<table>
<tr><th>x_i</th><th>n_i</th><th>p_i</th><th>n_i^2/p_i</th></tr>
<tr><td>0</td><td>4</td><td>0.057 4</td><td>278.745 7</td></tr>
<tr><td>1</td><td>8</td><td>0.154 1</td><td>390.006 1</td></tr>
<tr><td>2</td><td>14</td><td>0.234 4</td><td>836.177 5</td></tr>
<tr><td>3</td><td>19</td><td>0.223 3</td><td>1 616.659 2</td></tr>
<tr><td>4</td><td>10</td><td>0.159 5</td><td>629.959 2</td></tr>
<tr><td>5</td><td>4</td><td>0.091 1</td><td>175.631 2</td></tr>
<tr><td>6</td><td>2</td><td rowspan="4">0.070 2</td><td rowspan="4">227.920 2</td></tr>
<tr><td>7</td><td>1</td></tr>
<tr><td>8</td><td>1</td></tr>
<tr><td>≥9</td><td>0</td></tr>
<tr><td>Total</td><td>63</td><td>1</td><td>4155.099 1</td></tr>
</table>

It's easy to get

$$\chi^2=\frac{4\,155.099\,1}{63}-63\approx2.954\,0.$$

The number of group $l=7$, the number of unknown parameter $m=1$, so the degree of freedom of χ^2 is $k=l-m-1=5$. Checking the $\chi^2_\alpha(k)=\chi^2_{0.05}(5)=11.07$. For $\chi^2<\chi^2_{0.05}(5)$, so we should accept the null hypothesis, which is to say, the number of rain storm has a Poisson's distribution with $\hat{\lambda}\approx2.857\,1$.

8.6 Testing of Hypotheses Using p-value

A random sample $X_1, \cdots, X_{52}$ is drawn from a normal distribution with unknown μ and known σ^2 where $\sigma^2 = 100$. By calculating the means of this sample $\overline{x}=62.75$. We wish to test the hypothesis

$$H_0: \mu=\mu_0=60, \quad H_1: \mu>60.$$

We use the Z test. The test statistics is $Z=\dfrac{\overline{x}-\mu_0}{\sigma/\sqrt{n}}$. Therefore the observed value Z is $z_0=\dfrac{62.75-60}{10/\sqrt{52}}=1.983$.

The probability is $P(Z \geqslant z_0)=P(Z \geqslant 1.983)=1-\Phi(1.983)=0.0238$. This probability is called the p-value of right test of Z test. Notice that $P(Z \geqslant z_0)=p\text{-value}=0.0238$.

If the level of significance α is larger than $p=0.0238$, the corresponding critical value is $z_\alpha \leqslant 1.983$. That is to say the observed value $z_0=1.983$ is in the reject region. Therefore we should reject H_0. If the level of significance α is less than $p=0.0238$, the corresponding critical value is $z_\alpha \geqslant 1.983$. The observed value $z_0=1.983$ is not in the reject region. Therefore we should accept H_0.

So p-value$=P(Z \geqslant z_0)=0.0238$ is the least level of the significance to reject null-hypothesis H_0. In general, the definition of p-value is follows.

Definition 8.6.1

p-value of hypothesis test is the least level of the significance that the null-hypothesis should be rejected according observed sample value of testing statistics. (Fiure 8.6.1)

Example 8.6.1

The true average time to initial relief of pain for a best-selling pain relieveris known to be 10 min. Let μ denote the true average time to relief for a company's newly developed reliever. The company wishes to produce and market this reliever only if it provides quicker relief than the best-seller, so it wishes to test $H_0: \mu=10$, $H_1: \mu<10$. Only if experimental evidence leads to rejection of H_0 will the new reliever be introduced. After weighing the relative seriousness of each type of error, a single level of significance must be agreed on and a decision-to reject H_0 and introduce the reliever or not to do so-made at that level.

Suppose the new reliever has been introduced. The company supports its claim of quicker relief by stating that, based on an analysis of experimental data, $H_0: \mu=10$ was rejected in favor of $H_1: \mu<10$ using level of significance $\alpha=0.10$. Any individuals contemplating a switch to this new reliever would naturally want to reach their own conclusion concerning the validity of the claim. Individuals who are satisfied with the best-seller would view a type 1 error (concluding that the new product provides quicker relief when it actually does not) as serious, so might wish to use $\alpha=0.05$, 0.01, or even smaller levels. Unfortunately, the nature of the company's statement prevents an individual decision maker from reaching a conclusion at such a level. The company has imposed its own choice of significance level on others. The report could have been done in a manner that allowed each individual flexibility in drawing a conclusion at a personally selected α.

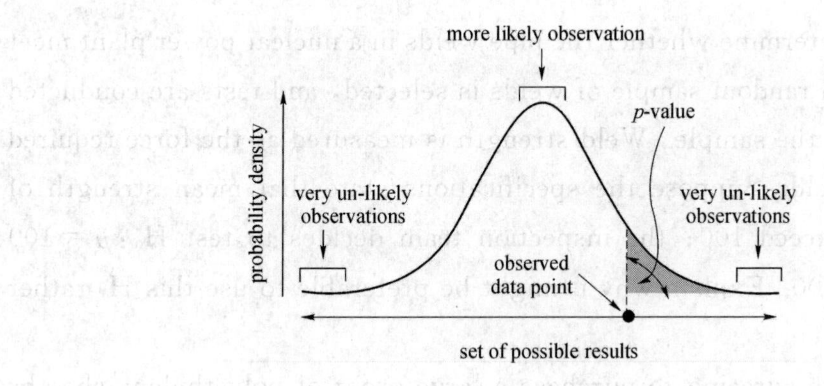

Figure 8. 6. 1

A *P*-value conveys much information about the strength of evidence against H_0 and allows an individual decision maker to draw a conclusion at any specified level α. Before we give a general definition, consider how the conclusion in a hypothesis-testing problem depends on the selected level α.

Example 8.6.2

There is a kind of elements with a life $X(h)$ having a distribution of $N(\mu,\sigma^2)$, the μ and σ^2 here are both unknown, and we measure the life of 16 elements, results are listed here:

159 280 101 212 224 379 179 264 222 362 168 250 149 260 485 170

Do we have sufficient reasons to say the average life of the element is greater than 225 h?

Here we will test $H_0:\mu\leqslant\mu_0=225, H_1:\mu>225$, let $\alpha=0.05$, it is easy to know the reject region of this test is:

$$t=\frac{\overline{x}-\mu_0}{s/\sqrt{n}}\geq t_\alpha(n-1)$$

and t_α here equals $t_\alpha=\dfrac{241.5-225}{98.7259/\sqrt{16}}=0.6685$, with the help of computer, we can calculate

$$p\text{-value}=P_{\mu_0}\{t\geqslant 0.6685\}=0.2570$$

p-value are greater than $\alpha=0.05$, so we will accept H_0.

Exercises

8.1 To determine whether the pipe welds in a nuclear power plant meet specifications, a random sample of welds is selected, and tests are conducted on each weld in the sample. Weld strength is measured as the force required to break the weld. Suppose the specifications state that mean strength of welds should exceed 100; the inspection team decides to test $H_0:\mu=100$ versus $H_1:\mu>100$. Explain why it might be preferable to use this H_1 rather than $\mu<100$.

8.2 Before agreeing to purchase a large order of polyethylene sheaths for a particular type of high pressure oil-filled submarine power cable, a company wants to see conclusive evidence that the true Standard deviation of sheath thickness is less than 0. 5mm. What hypotheses should be tested, and why? In this context, what are type 1 and type 2 errors?

8. 3 Let the test statistic Z have a standard normal distribution when H_0 is true. Give the significance level for each of the following situations.

(1) $H_1: \mu > \mu_0$ rejection region $z \geqslant 1.88$.

(2) $H_1: \mu < \mu_0$ rejection region $z \leqslant -2.75$.

(3) $H_1: \mu \neq \mu_0$ rejection region $z \leqslant -2.88$ or $z \geqslant 2.88$.

8. 4 The breaking strength of steel line produced by a factory has approximately normal distribution $X \sim N(\mu_0, 35^2)$ (unit: kg/cm²). A sample of 9 is selected and the sample average strength. $\bar{x}$ is 20 kg/cm² more than prior mean μ_0. If the standard deviation is not changed, can we consider the breaking strength of this kind steel line have increased significantly? (The significant level $\alpha = 0.05$)

8. 5 The amount of shaft wear (0. 000 1 in.) after a fixed mileage was determined for each of $n = 8$ internal combustion engines having copper lead as a bearing material, resulting in $\bar{x} = 3.72$ and $s = 1.25$. Assuming that the distribution of shaft wear is normal with mean μ, use the t test at level 0. 05 to test $H_0: \mu = 3.50$; $H_1: \mu > 3.50$.

8. 6 People found that when we brewing beer, it may produce NDMA in the malt drying. In the 1980 s, people discover a new way to dry the malt. Here are a list of the amount of NDMA in the old and new process:

New	2	1	2	2	1	0
Old	6	4	5	5	6	5
New	3	2	1	0	1	3
Old	5	6	4	6	7	4

Assuming they are all from a normal distribution and has same variance, let the mean of the old/new process be μ_1 / μ_2, we want to test $H_0: \mu_2 - \mu_1 \leqslant 2$; $H_1: \mu_2 - \mu_1 > 2 (\alpha = 0.05)$.

8. 7 A sample of 12 radon detectors of a certain type was selected, and

each was exposed to 100 pCi/L of radon. The resulting readings were as follows:

105.6 90.9 91.2 96.9 96.5 91.3 100.1

105.0 99.6 107.7 103.3 92.4

Does the data suggest that the population mean reading under these conditions differs from 100? State and test the appropriate hypotheses using $\alpha=0.05$.

8.8 We list the Growth days of two kinds of wheat here:

x	101	100	99	99	98	100	98	99	99	99
y	100	98	100	99	98	99	98	98	99	100

The two samples are independently selected from a normal distribution $N(\mu_1,\sigma_1^2)$ and $N(\mu_2,\sigma_2^2)$, here μ_i $i=1,2$ and σ_i^2 $i=1,2$ are all unknown. Test: $H_0:\sigma_1^2=\sigma_2^2$ $H_1:\sigma_1^2\neq\sigma_2^2(\alpha=0.05)$

8.9 There is a kind of mixed wheat with standard deviation of plant height $\sigma=14$cm, here we got a sample of 10, their plant height are:

90 105 101 95 100 100 101 105 93 97

Suppose theplant height obeys $N(\mu,\sigma^2)$, $\alpha=0.05$, we want to test if the standard deviation are bigger than the early group.

8.10 There are two machine production component. We get two sample with $n_1=60$ and $n_2=40$. Their sample variance are $s_1^2=15.46$ and $s_2^2=9.66$. The two samples are independently selected from a normal distribution $N(\mu_1,\sigma_1^2)$ and $N(\mu_2,\sigma_2^2)$, here μ_i $i=1,2$ and σ_i^2 $i=1,2$ are all unknown. We want to know: $(\alpha=0.05)$

$$H_0:\sigma_1^2\leqslant\sigma_2^2,\quad H_1:\sigma_1^2>\sigma_2^2.$$

8.11 Describe the difference between the mean of 95% confidence level and a significant level $\alpha=0.05$.

8.12 Here is a sample of students' grade of mathematics with $n=200$:

Grade x	20~30	30~40	40~50	50~60
number	5	15	30	51

Grade x	60～70	70～80	80～90	90～100
number	60	23	10	6

(1) Draw a histogram of the data.

(2) Test whether the data obeysnormal distribution $N(60,15^2)(\alpha=0.1)$.

8.13 A sample of 300 is selected from a batch of bulb, their working life are listing here:

t(hours)	0～100	100～200	200～300	>300
numbers	121	78	43	58

Test the hypotheses:

H_0: life of the bulb obeys:

$$f(t)=\begin{cases}0.005e^{-0.005t}, & t>0;\\ 0, & t\leqslant 0.\end{cases} \quad (\alpha=0.05)$$

8.14 Pairs of p-values and significance levels α are given. For each pair, state whether the observed p-value would lead to rejection of H_0 at given significance level.

(1) p-value$=0.084, \alpha=0.05$;

(2) p-value$=0.003, \alpha=0.001$;

(3) p-value$=0.498, \alpha=0.5$.

8.15 An aspirin manufacturer fills bottles by weight rather than by count. Since bottle should contain 100 tablet, the average weight per tablet should be 5 grains. Each of 100 tablets taken from a very large lot is weighted, resulting in a sample average weight per tablet of 4.87 grains and a sample standard deviation of 0.35 grain. Does the information provide strong evidence for concluding that the company is not filling its bottles as advertised? Test the appropriate hypotheses using $\alpha = 0.01$ by first computing the p-value and then comparing it to the specified significance level.

Chapter 9 Simple Linear Regression

9.1 the Method of Regression

Instatistics, regression is a statistical process for estimating the relationships among variables. It includes many techniques for modeling and analyzing several variables, when the focus is on the relationship between a dependent variable and one or more independent variables. More specifically, regression helps one understand how the typical value of the dependent variable (or "criterion variable") changes when any one of the independent variables is varied, while the other independent variables are held fixed. Most commonly, regression estimates the conditional expectation of the dependent variable given the independent variables-that is, the average value of the dependent variable when the independent variables are fixed. Less commonly, the focus is on a quantile, or other location parameter of the conditional distribution of the dependent variable given the independent variables. In all cases, the estimation target is a function of the independent variables called the regression function. In regression, it is also of interest to characterize the variation of the dependent variable around the regression function which can be described by a probability distribution.

Regression is widely used forprediction and forecasting, where its use has substantial overlap with the field of machine learning. Regression is also used to understand which among the independent variables are related to the dependent variable, and to explore the forms of these relationships. In restricted circumstances, regression can be used to infer causal relationships between the independent and dependent variables.

9.1.1 The method of Least Squares

The method of least squares is a standard approach to the approximate solution of regression model for constructing a predictor of one of the variables from the other by making use of a sample of observed pairs.

The most important application is indata fitting. The best fit in the least-squares sense minimizes the sum of squared residuals, a residual being the difference between an observed value and the fitted value provided by a model. When the problem has substantial uncertainties in the independent variable (the x variable), then simple regression and least squares methods have problems; in such cases, the methodology required for fitting errors-in-variables models may be considered instead of that for least squares.

9.1.2 Fitting a Straight Line

Table 9.1.1 Boiling point of water in degrees Fahrenheit and atmospheric pressure in inches of mercury from Forbes's experiments. These data are taken from Weisberg (1985, p.3).

Table 9.1.1

Boiling Point	Pressure
194.5	20.79
194.3	20.79
197.9	22.40
198.4	22.67
199.4	23.15
199.9	23.35
200.9	23.89
201.1	23.99
201.4	24.02
201.3	24.01
203.6	25.14
204.6	26.57

续表

Boiling Point	Pressure
209.5	28.49
208.6	27.76
210.7	29.04
211.9	29.88
212.2	30.06

Example 9.1.1 (Pressure and Boiling Point of Water)

Forbes (1857) reports the results from experiments that were trying to obtain a method for estimating altitude. A formula is available for altitude in terms of barometric pressure, but it was difficult to carry a barometer to high altitudes in Forbes' day. However, it might be easy for travelers to carry a thermometer and measure the boiling point of water. Table 9.1.1 contains the measured barometric pressures and boiling points of water from 17 experiments. We can use simple linear regression to estimate the relationship between boiling point and pressure. Let Y_i be the pressure for one of Forbes's observations, and let x_i be the corresponding boiling point for $i=1,\cdots,17$.

Figure 9.1.1 shows a plot of the observed values in Table 9.1.1.

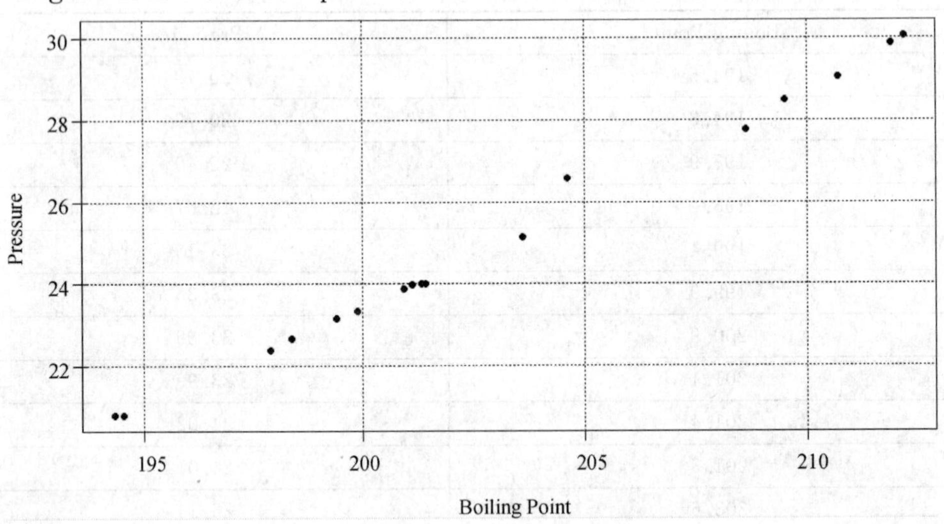

Figure 9.1.1

Suppose now that we are interested in describing the relationship between the Pressure y and boiling point x. In order to obtain a simple expression for this relationship, we might wish to fit a straight line to the 17 points plotted in Figure 9. 1. 1. Although these 17 points obviously do not lie exactly on a straight line, we might believe that the deviations from such a line are caused by the fact that the observed change under different conditions is affected not only by the weather but also by various other factors. In other words, we might believe that if it were possible to control all these other factors, the observed points would actually lie on a straight line. We might believe further that if we measured the pressure to the boiling point for a very large number of samples, instead of for just 17 experiments, we would then find that the observed points tend to cluster along a straight line. Perhaps we might also wish to be able to predict the pressure y of a unknown boiling point on the basis y in this experiment to the recorded boiling points x. One procedure for making such a prediction would be to fit a straight line to the points in Figure 9. 1. 1, and to use this line for predicting the value of y corresponding to each value of x.

9. 1. 3 The Least-Squares Line

We shall assume here that we are interested in fitting a straight Line to the points plotted in Figure 9. 1. 1 in order to obtain a simple mathematical relationship for expressing the pressure y to the boiling point x. In other words, our main objective is to be able to predict closely pressure y from its boiling point x. We are interested, therefore, in constructing a straight line such that, for each observed boiling point x_i, the corresponding value of y on the straight line will be as close as possible to the actual observed response y_i.

The method of least squares is one method of constructing a straight line to fit the observed values. The line should be drawn, making the sum of the squares of the vertical deviations of all the points from the line is a minimum, according to this method. Then, we shall learn this method in detail.

Figure 9.1.2 shows a straight line fitted to all poins in Table 9.1.1.

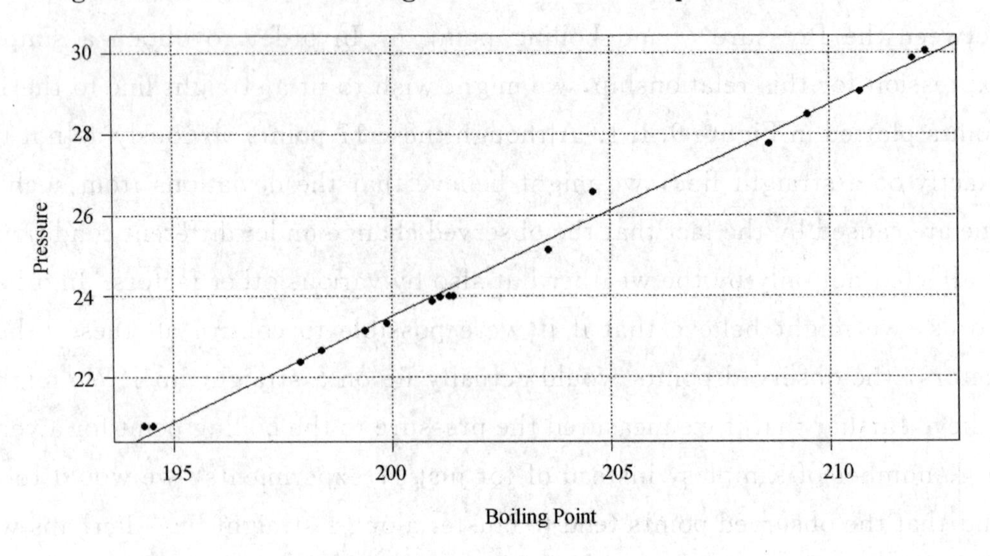

Figure 9.1.2

Think about an arbitrary straight line $y=\beta_0+\beta_1 x$, where the values of the constants β_0 and β_1 are to be determined. The height of this line is $\beta_0+\beta_1 x_i$ when $x=x_i$. Therefore, the distance between the point (x_i, y_i) and the line is $|y_i-(\beta_0+\beta_1 x_i)|$. Suppose that we use these n points to fit the line $\beta_0+\beta_1 x_i$, and let Q denote the sum of the squares of the vertical distances at the n points. Then

$$Q=\sum_{i=1}^{n}[y_i-(\beta_0+\beta_1 x_i)]^2 \tag{9.1.1}$$

The method of least squaresassigns that the values of β_0 and β_1 must be chosen to minimize the value of Q.

The detailed procedure will be described in the section 9.2. Here just show the result of least square estimates.

For the values given in Table 9.1.1, $n=17$, $\hat{\beta}_0=-81.063\,73$ and $\hat{\beta}_1=0.522\,89$. Hence, the equation of the least-squares line is $y=-81.063\,73+0.522\,89x$. This line is sketched in Figure 9.1.2.

Almost all of the statistical computer software will calculate the least squares regression line. Even some of the hand-held calculator to do the calculation.

9.1.4 Fitting a Polynomial by the Method of Least Squares

Suppose now that we wish to fit a polynomial of degree $k(k \geqslant 2)$, rather than simply fitting a straight line to n plotted points. Such a polynomial will have the following form:

$$y=\beta_0+\beta_1 x+\beta_2 x^2+\cdots+\beta_k x^k.$$

Without loss of generality, we assume that the degree of polynomial is less than the sample size. The value of the constants $\beta_0, \cdots, \beta_k$ should be satisfied the minimum sum Q of the squares of the difference of the points from the curve. That is, these constants should be selected so that they could minimize the following expression for Q:

$$Q=\sum_{i=1}^{n}[y_i-(\beta_0+\beta_1 x_i+\cdots+\beta_k x_i^k)]^2. \tag{9.1.2}$$

Here we give roughly steps. Calculating the $k+1$ partial derivatives $\partial Q/\partial\beta_0, \cdots, \partial Q/\partial\beta_k$, and setting each of these derivatives equal to 0, we obtain the following $k+1$ linear equations involving the $k+1$ unknown values $\beta_0, \cdots, \beta_k$:

$$\begin{aligned}
\beta_0 n+\beta_1 \sum_{i=1}^{n} x_i+\cdots+\beta_k \sum_{i=1}^{n} x_i^k &= \sum_{i=1}^{n} y_i; \\
\beta_0 \sum_{i=1}^{n} x_i+\beta_1 \sum_{i=1}^{n} x_i^2+\cdots+\beta_k \sum_{i=1}^{n} x_i^{k+1} &= \sum_{i=1}^{n} x_i y_i; \\
&\vdots \\
\beta_0 \sum_{i=1}^{n} x_i^k+\beta_1 \sum_{i=1}^{n} x_i^{k+1}+\cdots+\beta_k \sum_{i=1}^{n} x_i^{2k} &= \sum_{i=1}^{n} x_i^k y_i.
\end{aligned}$$

As previously mentioned, these equations are called the normal equations. If and only if the determinant of the $(k+1)\times(k+1)$ matrix formed from the coefficients of $\beta_0, \cdots, \beta_k$ is not zero, there will be a unique set of values of $\beta_0, \cdots, \beta_k$ which satisfy the normal equations. The determinant is not zero, and there is a unique solution of the normal equations while existing at least of $k+1$ different set of values in the observed values $x_1, \cdots, x_n$. We will assume the condition to be satisfied. The unique values of $\beta_0, \cdots, \beta_k$ can be shown by the methods of advanced calculus, satisfying the normal equations will then be the values which minimize the value of Q in Eq. (9.1.2). If we express these values by $\hat{\beta}_0, \cdots, \hat{\beta}_k$, then the least-squares polynomial is $y=\hat{\beta}_0+\hat{\beta}_1 x+\cdots+\hat{\beta}_k x^k$.

Example 9.1.2

The followingtable shows the average unit price of a product Y and the number x of a batch.

Table 9.1.2

x	20	25	30	35	40	50	60	65	70	75	80	90
y	1.81	1.70	1.65	1.55	1.48	1.40	1.30	1.26	1.24	1.21	1.20	1.18

Here we construct a polynomial of degree 2. That is we should fit a polynomial of the form $y=\beta_0+\beta_1 x+\beta_2 x^2$ to the 12 points given in Table 9.1.2. As the procedure in front of this example, the unique values of β_0, β_1 and β_2 are $\hat{\beta}_0=2.198\,266\,29$, $\hat{\beta}_1=-0.022\,522\,36$ and $\hat{\beta}_2=0.000\,125\,07$. Hence the least-squares parabola is

$$y=2.198\,266\,29-0.022\,522\,36x+0.000\,126\,07x^2.$$

9.2 Estimation and Inference in Simple Linear Regression

9.2.1 Estimation Parameters

Let $x_1, \cdots, x_n$ denote values of the independent variable for which observations are made, and let y_i denote observed value associated with x_i. The available data then consists of the n pairs $(x_1, Y_1), \cdots, (x_n, Y_n)$. Then

$$Y_i=a+bx_i+\varepsilon_i, \quad \varepsilon_i \sim N(0, \sigma^2).$$

Where ε_i are independent mutually.

Then $Y_i \sim N(a+bx_i, \sigma^2)$, $i=1, 2, \cdots, n$. Since $Y_1, \cdots, Y_n$ are independent, the joint distribution of $Y_1, \cdots, Y_n$ is

$$L = \prod_{i=1}^{n} \frac{1}{\sigma\sqrt{2\pi}} \exp\left[-\frac{1}{2\sigma^2}(y_i - a - bx_i)^2\right]$$

$$= \left(\frac{1}{\sigma\sqrt{2\pi}}\right)^n \exp\left[-\frac{1}{2\sigma^2}\sum_{i=1}^{n}(y_i - a - bx_i)^2\right].$$

Then use maximum likelihoodmethod to estimate unknown parameter a, b. The upper formula is the likelihood function of sample $y_1, \cdots, y_n$. Obviously, the maximum of this formula is equivalent to minimizing

$$Q(a, b) = \sum_{i=1}^{n}(y_i - a - bx_i)^2.$$

The minimizing values of a, b are found by taking partial derivatives of $Q(a, b)$ with respect to both a and b, equating them both to zero, and solving the equations

$$\begin{cases} \dfrac{\partial Q}{\partial a} = -2\sum\limits_{i=1}^{n}(y_i - a - bx_i) = 0; \\ \dfrac{\partial Q}{\partial b} = -2\sum\limits_{i=1}^{n}(y_i - a - bx_i)x_i = 0. \end{cases}$$

Cancellation of the 2 factor and rearrangement gives the following system of equations, called the **normal equations**:

$$\begin{cases} na + (\sum\limits_{i=1}^{n})b = \sum\limits_{i=1}^{n} y_i; \\ (\sum\limits_{i=1}^{n} x_i)a + (\sum\limits_{i=1}^{n} x_i^2)b = \sum\limits_{i=1}^{n} x_i y_i. \end{cases}$$

Because x_i are not all the same, the coefficient determinant of the normal equations

$$\begin{vmatrix} n & \sum\limits_{i=1}^{n} x_i \\ \sum\limits_{i=1}^{n} x_i & \sum\limits_{i=1}^{n} x_i^2 \end{vmatrix} = n\sum_{i=1}^{n} x_i^2 - (\sum_{i=1}^{n} x_i)^2 = n(\sum_{i=1}^{n} x_i - \bar{x})^2 \neq 0.$$

Therefore the least squares estimates are the unique solution to this system. The least square estimates of a and b are

$$\begin{cases} \hat{b} = \dfrac{n\sum_{i=1}^{n} x_i y_i - (\sum_{i=1}^{n} x_i)(\sum_{i=1}^{n} y_i)}{n\sum_{i=1}^{n} x_i^2 - (\sum_{i=1}^{n} x_i)^2} = \dfrac{\sum_{i=1}^{n}(x_i - \overline{x})(y_i - \overline{y})}{\sum_{i=1}^{n}(x_i - \overline{x})^2}; \\ \hat{a} = \dfrac{1}{n}\sum_{i=1}^{n} y_i - \dfrac{\hat{b}}{n}\sum_{i=1}^{n} x_i = \overline{y} - \hat{b}\overline{x}. \end{cases} \tag{9.2.1}$$

Where $\overline{x} = \frac{1}{n}\sum_{i=1}^{n} x_i, \overline{y} = \frac{1}{n}\sum_{i=1}^{n} y_i$.

For the given x, we treat $\hat{a} + \hat{b}x$ to be the estimation of regression function $\mu(x) = a + bx$. That is $\widehat{\mu(x)} = \hat{a} + \hat{b}x$, called Empirical regression equation of Y on x.

According to $\hat{a}$, the regression equation can be rewritten as

$$\hat{y} = \overline{y} + \hat{b}(x - \overline{x}).$$

From this expression, we know that the estimated regression line pass the geometric center $(\overline{x}, \overline{y})$. For convenience, we note that

$$\begin{cases} S_{xx} = \sum_{i=1}^{n}(x_i - \overline{x})^2 = \sum_{i=1}^{n} x_i^2 - \frac{1}{n}(\sum_{i=1}^{n} x_i)^2; \\ S_{yy} = \sum_{i=1}^{n}(y_i - \overline{y})^2 = \sum_{i=1}^{n} y_i^2 - \frac{1}{n}(\sum_{i=1}^{n} y_i)^2; \\ S_{xy} = \sum_{i=1}^{n}(x_i - \overline{x})(y_i - \overline{y}) = \sum_{i=1}^{n} x_i y_i - \frac{1}{n}(\sum_{i=1}^{n} x_i)(\sum_{i=1}^{n} y_i). \end{cases}$$

Thus the estimator of a, b can be written as

$$\begin{cases} \hat{b} = \dfrac{S_{xy}}{S_{xx}}; \\ \hat{a} = \dfrac{1}{n}\sum_{i=1}^{n} y_i - (\dfrac{1}{n}\sum_{i=1}^{n} x_i)\hat{b}. \end{cases} \tag{9.2.2}$$

Example 9.2.1 (Pressure and Boiling Point of Water)

Forbes was trying to use the boiling point of water to estimate the barometric pressure. Using the data in Table 9.1.1, we can compute the least-squares estimates of the simple linear regression.

Let Y_i be the pressure for one of Forbes's observations, and let x_i be the corresponding boiling point for $i=1, \cdots, 17$. We model the Y_i as being independent with means $\beta_0+\beta_1 x_i$ and variance σ^2. The average temperature is $\bar{x}=202.95$ and $s_x^2=530.78$ with $n=17$.

Then the regression equation is

$$\hat{y}=-81.0637+0.5229x.$$

9.2.2 Estimation of σ^2

From the regression equation

$$E\{[Y-(a+bx)]^2\}=E(\varepsilon^2)=D(\varepsilon^2)+[E(\varepsilon)]^2=\sigma^2,$$

we know that the smaller σ^2, the smaller mean square error of the regression function $\mu(x)=a+bx$. Thus we use the regression function $\mu(x)=a+bx$ to study the relation of random variable Y and x. But σ^2 are unknown. We should use sample to estimate σ^2.

Denote that $\hat{y}_i=\hat{y}\,|_{x=x_i}=\hat{a}+\hat{b}x_i$, $y_i-\hat{y}_i$ called the residual at x_i. The sum of square

$$Q_e=\sum_{i=1}^{n}(y_i-\hat{y}_i)^2=\sum_{i=1}^{n}(y_i-\hat{a}-\hat{b}x_i)^2,$$

which is called residual sum of squares. Then Q_e can be decomposed as follows:

$$\begin{aligned} Q_e &= \sum_{i=1}^{n}(y_i-\hat{y}_i)^2=\sum_{i=1}^{n}[y_i-\bar{y}-\hat{b}(x_i-\bar{x})]^2 \\ &= \sum_{i=1}^{n}(y_i-\bar{y})^2-2\hat{b}\sum_{i=1}^{n}(x_i-\bar{x})(y_i-\bar{y})+\hat{b}^2\sum_{i=1}^{n}(x_i-\bar{x})^2 \\ &= S_{yy}-2\hat{b}S_{xy}+\hat{b}^2S_{xx}. \end{aligned}$$

From $\hat{b}=S_{xy}/S_{xx}$,

$$Q_e=S_{yy}-\hat{b}S_{xy}.$$

The distribution of Q_e is

$$\frac{Q_e}{\sigma^2} \sim \chi^2(n-2).$$

Then $E\left(\frac{Q_e}{\sigma^2}\right) = n - 2$. That is $E(Q_e/(n-2)) = \sigma^2$. So the unbiased estimator of σ^2 is as follows:

$$\hat{\sigma}^2 = \frac{Q_e}{n-2} = \frac{1}{n-2}(S_{yy} - \hat{b}S_{xy}). \qquad (9.2.3)$$

Example 9.2.2

The unbiased estimator of σ^2 in **Example 9.1.1.**

$$S_{yy} = \sum_{i=1}^{n} y_i^2 - \frac{1}{n}\left(\sum_{i=1}^{n} y_i\right)^2 = 10\,821 - \frac{1}{17} \times 426^2 = 145.937\,8,$$

$$S_{xy} = \sum_{i=1}^{n} (y_i - \bar{y})(x_i - \bar{x}) = 277.542\,1,$$

$$\hat{b} = 0.522\,89.$$

Therefore

$$Q_e = S_{yy} - \hat{b}S_{xy} = 0.813\,8,$$

$$\hat{\sigma}^2 = 0.0542.$$

9.2.3 Tests of Hypotheses of linear assumption

In the former discussion, we assume thatthe regression of Y on x has the formula like $a + bx$. In general, we should test the hypothesis of the regression coefficient.

Therefore we need test the hypotheses

$$H_0: b = 0.$$

$$H_1: b \neq 0.$$

We use t statistics to test,

$$\hat{b} \sim N(b, \sigma^2/S_{xx}).$$

Then

$$\frac{(n-2)\hat{\sigma}^2}{\sigma^2} = \frac{Q_e}{\sigma^2} \sim \chi^2(n-2).$$

Since $\hat{b}$ and Q_e are independent,

$$\frac{\hat{b}-b}{\sqrt{\sigma^2/S_{xx}}}\Big/\sqrt{\frac{(n-2)\hat{\sigma}^2}{\sigma^2}/(n-2)}\sim t(n-2).$$

That is

$$\frac{\hat{b}-b}{\hat{\sigma}}\sqrt{S_{xx}}\sim t(n-2).$$

Where $\hat{\sigma}=\sqrt{\hat{\sigma}^2}$.

When H_0 is true,

$$t=\frac{\hat{b}}{\hat{\sigma}}\sqrt{S_{xx}}\sim t(n-2).$$

And $E(\hat{b})=b=0$. So the region of reject is

$$|t|=\frac{|\hat{b}|}{\hat{\sigma}}\sqrt{S_{xx}}\geqslant t_{\alpha/2}(n-2). \qquad (9.2.4)$$

Where α is the significance level.

Example 9.2.3

To test whether the regression coefficient is significant in Example 9.1.1, let $\alpha=0.05$.

Since $\alpha=0.05$, our critical value will be the $0.05/2=0.025$ quantile of the t distribution with 17 degrees of freedom. From Example 9.2.3, we know that $\hat{b}=0.522\,89$, $\hat{\sigma}^2=0.054\,2$, $S_{xx}=530.784\,9$. From the Table of t distribution in this book, we find that $t_{0.05/n}(n-2)=t_{0.025}(15)=2.131\,5$. So the t test specifies rejecting H_0 if $|t|=\frac{|\hat{b}|}{\hat{\sigma}}\sqrt{S_{xx}}\geqslant 2.131\,5$. Since $|t|=\frac{0.522\,89}{\sqrt{0.054\,2}}\sqrt{530.784\,9}=51.745\,19\geqslant 2.131\,5$, the hypothesis H_0 would be rejected.

9.2.4 Confidence interval of parameter b

When accept H_0, we should give a confidence interval for $\hat{b}$ with confidence coefficient $1-\alpha$ as follows:

$$(\hat{b}\pm t_{\alpha/2}(n-2)\times\frac{\hat{\sigma}}{\sqrt{S_{xx}}}).\tag{9.2.5}$$

The confidence interval with 0.95 in Example 9.1.1 is

$$(0.522\,89\pm 2.131\,5\times\sqrt{\frac{0.054\,2}{530.784\,9}})=(0.501\,35, 0.544\,43).$$

9.2.5 Point estimation and confidence interval of regression function

Treat $\hat{y}=\widehat{\mu(x)}=\hat{a}+\hat{b}x$ is the point estimator of $\mu(x_0)=a+bx_0$, that is

$$\hat{y}_0=\widehat{\mu(x_0)}=\hat{a}+\hat{b}x_0.$$

Considering the corresponding estimator

$$\hat{Y}_0=\hat{a}+\hat{b}x_0.$$

Since this estimator is unbiased, then

$$\frac{\hat{Y}_0-(a+bx_0)}{\sigma\sqrt{\frac{1}{n}+\frac{(x_0-\bar{x})^2}{S_{xx}}}}\sim N(0,1).$$

Because

$$\frac{(n-2)\hat{\sigma}^2}{\sigma^2}=\frac{Q_e}{\sigma^2}\sim\chi^2(n-2),$$

Q_e and $\hat{Y}_0$ are independent. Then

$$\frac{\hat{Y}_0-(a+bx_0)}{\sigma\sqrt{\frac{1}{n}+\frac{(x_0-\bar{x})^2}{S_{xx}}}}\Bigg/\sqrt{\frac{(n-2)\hat{\sigma}^2}{\sigma^2}/(n-2)}\sim t(n-2).$$

Thenconfidence interval of $\mu(x_0)=a+bx_0$ with confidence coefficient $1-\alpha$ is

$$(\hat{Y}_0\pm t_{\alpha/2}(n-2)\hat{\sigma}\sqrt{\frac{1}{n}+\frac{(x_0-\bar{x})^2}{S_{xx}}})\tag{9.2.6}$$

or

$$(\hat{a}+\hat{b}x_0 \pm t_{a/2}(n-2)\hat{\sigma}\sqrt{\frac{1}{n}+\frac{(x_0-\overline{x})^2}{S_{xx}}}). \tag{9.2.7}$$

This length of confidence interval is function of x_0, it increases as $|x_0-\overline{x}|$ increases.

9.2.6 Point estimator and confidence interval of prediction

Presume that n pairs observations $(x_1,Y_1),\cdots,(x_n,Y_n)$ are to be gotten from a problem of simple linear regression. The value of an independent observation Y_0 can be predicted on the basis of these n pairs, which will be obtained when a certain specified value x is assigned to the control variable. It is natural to use the value $\hat{Y}=\hat{\beta}_0+\hat{\beta}_1 x$ as the predicted value of the observation Y, because Y will have a normal distribution with mean $\beta_0+\beta_1 x$ and variance σ^2. Now, we will determine the M. S. E. $E[(\hat{Y}-Y)^2]$ of this prediction, in which both $\hat{Y}$ and Y are random variables.

In this problem, $E[\hat{Y}]=E[Y]=\beta_0+\beta_1 x$. Thus, if we let $\mu=\beta_0+\beta_1 x$, then

$$\begin{aligned}E[(\hat{Y}-Y)^2]&=E\{[(\hat{Y}-\mu)-(Y-\mu)]^2\}\\&=\mathrm{Var}(\hat{Y})+\mathrm{Var}(Y)-2\mathrm{Cov}(\hat{Y},Y).\end{aligned}$$

However, the random variable $\hat{Y}$ is a function of the first n pairs of observations and the random variable Y is an independent observation so that they are independent.

Therefore, $\mathrm{Cov}(\hat{Y},Y)=0$, and it follows that

$$E[(\hat{Y}-Y)^2]=\mathrm{Var}(\hat{Y})+\mathrm{Var}(Y).$$

At last, the variance of $\hat{Y}=\hat{\beta}_0+\hat{\beta}_1 x$ is equal to $\sigma^2\left(\frac{1}{n}+\frac{(x-\overline{x})^2}{S_{xx}}\right)$, Owing to $\mathrm{Var}(Y)=\sigma^2$, we have

$$E[(\hat{Y}-Y)^2]=\sigma^2\left[1+\frac{1}{n}+\frac{(x-\overline{x})^2}{S_{xx}}\right].$$

Therefore

$$\hat{Y}-Y\sim N\left(0,\left[1+\frac{1}{n}+\frac{(x-\bar{x})^2}{S_{xx}}\right]\sigma^2\right).$$

That is

$$\frac{\hat{Y}-Y}{\sigma\sqrt{1+\frac{1}{n}+\frac{(x-x)^2}{S_{xx}}}}\sim N(0,1). \tag{9.2.8}$$

According to the independence,

$$\frac{\hat{Y}-Y}{\sigma\sqrt{1+\frac{1}{n}+\frac{(x-\bar{x})^2}{S_{xx}}}}\Bigg/\sqrt{\frac{(n-2)\hat{\sigma^2}}{\sigma^2}/(n-2)}\sim t(n-2).$$

That is

$$\frac{\hat{Y}-Y}{\hat{\sigma}\sqrt{1+\frac{1}{n}+\frac{(x-\bar{x})^2}{S_{xx}}}}\sim t(n-2).$$

For confidenceinterval with confidence coefficient $1-\alpha$, we have

$$P\left(\frac{|\hat{Y}-Y|}{\hat{\sigma}\sqrt{1+\frac{1}{n}+\frac{(x-\bar{x})^2}{S_{xx}}}}\sim t_{\alpha/2}(n-2)\right)=1-\alpha$$

or

$$P\left[\hat{Y}-t_{\alpha/2}(n-2)\hat{\sigma}\sqrt{1+\frac{1}{n}+\frac{(x-\bar{x})^2}{S_{xx}}}<Y\right.$$

$$\left.<\hat{Y}+t_{\alpha/2}(n-2)\hat{\sigma}\sqrt{1+\frac{1}{n}+\frac{(x-\bar{x})^2}{S_{xx}}}\right]=1-\alpha.$$

Interval

$$\left(\hat{Y}\pm t_{\alpha/2}(n-2)\hat{\sigma}\sqrt{1+\frac{1}{n}+\frac{(x-\bar{x})^2}{S_{xx}}}\right).$$

That is

$$\left(\hat{\beta}_0+\hat{\beta}_1x+t_{\alpha/2}(n-2)\hat{\sigma}\sqrt{1+\frac{1}{n}+\frac{(x_0-\bar{x})^2}{S_{xx}}}\right). \tag{9.2.9}$$

Called the prediction interval with confidence coefficient $1-\alpha$.

Example 9.2.5

For the data present in Table 9.1.1, construct a confidence interval with confidence coefficient 0.95 for the height of the regression line at the point $x_0=206$.

From Example 9.2.1, Example 9.2.3, and Example 9.2.4, we have $\hat{b}=0.522\,89$, $\hat{a}=-81.063\,7$, $S_{xx}=530.78$, $\hat{\sigma}^2=0.054\,2$, and $\bar{x}=202.95$. From the Table of t distribution in this book, we find that $t_{0.05/2}(15)=2.131$.

$$\hat{Y}_0=\hat{Y}|_{x_0=206}=[-81.063\,7+0.522\,89x_0]|_{x_0=206}=26.651\,64,$$

$$t_{\alpha/2}(n-2)\hat{\sigma}\sqrt{\frac{1}{n}+\frac{(x_0-\bar{x})}{S_{xx}}}=2.131\times\sqrt{0.054\,2}\times\sqrt{1+\frac{1}{17}+\frac{(206-202.95)^2}{530.78}}$$
$$=0.5147.$$

Therefore the confidence interval with confidence coefficient 0.95 at the point $x_0=206$ is

$$(26.651\,64\pm0.514\,7).$$

Exercise

9.1 Prove that $\sum_{i=1}^{n}(c_1x_i+c_2)^2=c_1\sum_{i=1}^{n}(x_i-\bar{x})^2+n(c_1\bar{x}+c_2)^2$.

9.2 Show that the value of $\hat{\beta}_1$ can be rewritten in each of the following forms:

(a) $\hat{\beta}_1=\dfrac{\sum_{i=1}^{n}(x_i-\bar{x})(y_i-\bar{y})}{\sum_{i=1}^{n}(x_i-\bar{x})^2}$.

(b) $\hat{\beta}_1=\dfrac{\sum_{i=1}^{n}(x_i-\bar{x})y_i}{\sum_{i=1}^{n}(x_i-\bar{x})^2}$.

9.3 Show that the least-squares line $y=\hat{\beta}_0+\hat{\beta}_1x$ passes through the point $(\bar{x},\bar{y})$.

9.4 Suppose that eight specimens of a certain type of alloy were produced at different temperatures, and the durability of each specimen was then observed. The observed values are given in the following Table, where x_i denotes the temperature (in coded units) at which specimen I was produced and y_i denotes durability (in coded units) of that specimen.

i	x_i	y_i
1	0.5	40
2	1.0	41
3	1.5	43
4	2.0	42
5	2.5	44
6	3.0	42
7	3.5	43
8	4.0	42

(1) Fit a straight line of the form $y=\beta_0+\beta_1 x$ to these values by the method of least squares.

(2) Fit a parabola of the form $y=\beta_0+\beta_1 x+\beta_2 x^2$ to these values by the method of least squares.

9.5 Show that $E(\hat{\beta}_1)=\beta_1$.

9.6 Show that $E(\hat{\beta}_0)=\beta_0$.

9.7 Show that $\operatorname{Var}(\hat{\beta}_0)=\sigma^2\left(\frac{1}{n}+\frac{\overline{x}^2}{s_x^2}\right)$.

9.8 Show that $\operatorname{Cov}(\hat{\beta}_0,\hat{\beta}_1)=-\frac{\overline{x}\sigma^2}{s_x^2}$.

9.9 Consider a problem of simple linear regression in which the durability Y of a certain type of alloy is to be related to the temperature X at which it was produced. Suppose that eight pairs of observed values given in Question 9.4 are obtained. Determine the values of the M. L. E.'s $\hat{\beta}_0$, $\hat{\beta}_1$, and $\hat{\sigma}^2$, and also the values of $\mathrm{Var}(\hat{\beta}_0)$ and $\mathrm{Var}(\hat{\beta}_1)$.

9.10 Consider again the conditions of Exercise 9. If a specimen of the alloy is to be produced at the temperature $x=3.5$, what is the predicted value of the durability of the specimen and construct a confidence interval with confidence coefficient 0.95.

9.11 Consider again the conditions of Exercise 9. If a specimen of the alloy is to be produced at new temperature $x=4.25$, what is the predicted value of the durability of the specimen and construct a confidence interval with confidence coefficient 0.95.

Table 1 Table of Standard Normal Distribution Function

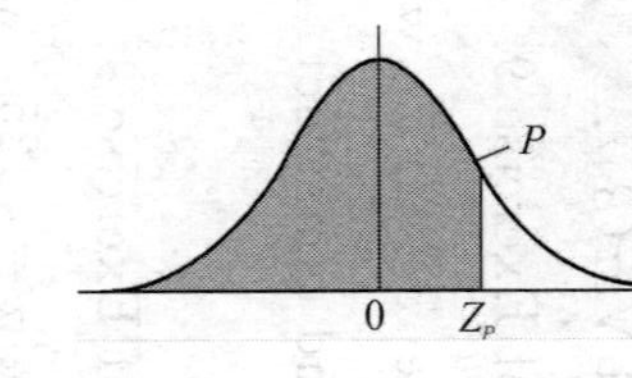

$$\Phi(x)=\int_{-\infty}^{x}\frac{1}{\sqrt{2\pi}}e^{-\frac{x^2}{2}}dx$$

x	0.00	0.01	0.02	0.03	0.04	0.05	0.06	0.07	0.08	0.09
0.00	0.500 000	0.503 989	0.507 978	0.511 966	0.515 953	0.519 939	0.523 922	0.527 903	0.531 881	0.535 856
0.10	0.539 828	0.543 795	0.547 758	0.551 717	0.555 670	0.559 618	0.563 559	0.567 495	0.571 424	0.575 345
0.20	0.579 260	0.583 166	0.587 064	0.590 954	0.594 835	0.598 706	0.602 568	0.606 420	0.610 261	0.614 092
0.30	0.617 911	0.621 720	0.625 516	0.629 300	0.633 072	0.636 831	0.640 576	0.644 309	0.648 027	0.651 732
0.40	0.655 422	0.659 097	0.662 757	0.666 402	0.670 031	0.673 645	0.677 242	0.680 822	0.684 386	0.687 933
0.50	0.691 462	0.694 974	0.698 468	0.701 944	0.705 401	0.708 840	0.712 260	0.715 661	0.719 043	0.722 405
0.60	0.725 747	0.729 069	0.732 371	0.735 653	0.738 914	0.742 154	0.745 373	0.748 571	0.751 748	0.754 903
0.70	0.758 036	0.761 148	0.764 238	0.767 305	0.770 350	0.773 373	0.776 373	0.779 350	0.782 305	0.785 236
0.80	0.788 145	0.791 030	0.793 892	0.796 731	0.799 546	0.802 337	0.805 105	0.807 850	0.810 570	0.813 267

续表

x	0.00	0.01	0.02	0.03	0.04	0.05	0.06	0.07	0.08	0.09
0.90	0.815 940	0.818 589	0.821 214	0.823 814	0.826 391	0.828 944	0.831 472	0.833 977	0.836 457	0.838 913
1.00	0.841 345	0.843 752	0.846 136	0.848 495	0.850 830	0.853 141	0.855 428	0.857 690	0.859 929	0.862 143
1.10	0.864 334	0.866 500	0.868 643	0.870 762	0.872 857	0.874 928	0.876 976	0.879 000	0.881 000	0.882 977
1.20	0.884 930	0.886 861	0.888 768	0.890 651	0.892 512	0.894 350	0.896 165	0.897 958	0.899 727	0.901 475
1.30	0.903 200	0.904 902	0.906 582	0.908 241	0.909 877	0.911 492	0.913 085	0.914 657	0.916 207	0.917 736
1.40	0.919 243	0.920 730	0.922 196	0.923 641	0.925 066	0.926 471	0.927 855	0.929 219	0.930 563	0.931 888
1.50	0.933 193	0.934 478	0.935 745	0.936 992	0.938 220	0.939 429	0.940 620	0.941 792	0.942 947	0.944 083
1.60	0.945 201	0.946 301	0.947 384	0.948 449	0.949 497	0.950 529	0.951 543	0.952 540	0.953 521	0.954 486
1.70	0.955 435	0.956 367	0.957 284	0.958 185	0.959 070	0.959 941	0.960 796	0.961 636	0.962 462	0.963 273
1.80	0.964 070	0.964 852	0.965 620	0.966 375	0.967 116	0.967 843	0.968 557	0.969 258	0.969 946	0.970 621
1.90	0.971 283	0.971 933	0.972 571	0.973 197	0.973 810	0.974 412	0.975 002	0.975 581	0.976 148	0.976 705
2.00	0.977 250	0.977 784	0.978 308	0.978 822	0.979 325	0.979 818	0.980 301	0.980 774	0.981 237	0.981 691
2.10	0.982 136	0.982 571	0.982 997	0.983 414	0.983 823	0.984 222	0.984 614	0.984 997	0.985 371	0.985 738
2.20	0.986 097	0.986 447	0.986 791	0.987 126	0.987 455	0.987 776	0.988 089	0.988 396	0.988 696	0.988 989
2.30	0.989 276	0.989 556	0.989 830	0.990 097	0.990 358	0.990 613	0.990 863	0.991 106	0.991 344	0.991 576
2.40	0.991 802	0.992 024	0.992 240	0.992 451	0.992 656	0.992 857	0.993 053	0.993 244	0.993 431	0.993 613
2.50	0.993 790	0.993 963	0.994 132	0.994 297	0.994 457	0.994 614	0.994 766	0.994 915	0.995 060	0.995 201

续表

x	0.00	0.01	0.02	0.03	0.04	0.05	0.06	0.07	0.08	0.09
2.60	0.995 339	0.995 473	0.995 604	0.995 731	0.995 855	0.995 975	0.996 093	0.996 207	0.996 319	0.996 427
2.70	0.996 533	0.996 636	0.996 736	0.996 833	0.996 928	0.997 020	0.997 110	0.997 197	0.997 282	0.997 365
2.80	0.997 445	0.997 523	0.997 599	0.997 673	0.997 744	0.997 814	0.997 882	0.997 948	0.998 012	0.998 074
2.90	0.998 134	0.998 193	0.998 250	0.998 305	0.998 359	0.998 411	0.998 462	0.998 511	0.998 559	0.998 605
3.00	0.998 650	0.998 694	0.998 736	0.998 777	0.998 817	0.998 856	0.998 893	0.998 930	0.998 965	0.998 999
3.10	0.999 032	0.999 065	0.999 096	0.999 126	0.999 155	0.999 184	0.999 211	0.999 238	0.999 264	0.999 289
3.20	0.999 313	0.999 336	0.999 359	0.999 381	0.999 402	0.999 423	0.999 443	0.999 462	0.999 481	0.999 499
3.30	0.999 517	0.999 534	0.999 550	0.999 566	0.999 581	0.999 596	0.999 610	0.999 624	0.999 638	0.999 651
3.40	0.999 663	0.999 675	0.999 687	0.999 698	0.999 709	0.999 720	0.999 730	0.999 740	0.999 749	0.999 758
3.50	0.999 767	0.999 776	0.999 784	0.999 792	0.999 800	0.999 807	0.999 815	0.999 822	0.999 828	0.999 835
3.60	0.999 841	0.999 847	0.999 853	0.999 858	0.999 864	0.999 869	0.999 874	0.999 879	0.999 883	0.999 888
3.70	0.999 892	0.999 896	0.999 900	0.999 904	0.999 908	0.999 912	0.999 915	0.999 918	0.999 922	0.999 925
3.80	0.999 928	0.999 931	0.999 933	0.999 936	0.999 938	0.999 941	0.999 943	0.999 946	0.999 948	0.999 950
3.90	0.999 952	0.999 954	0.999 956	0.999 958	0.999 959	0.999 961	0.999 963	0.999 964	0.999 966	0.999 967
4.00	0.999 968	0.999 970	0.999 971	0.999 972	0.999 973	0.999 974	0.999 975	0.999 976	0.999 977	0.999 978
4.10	0.999 979	0.999 980	0.999 981	0.999 982	0.999 983	0.999 983	0.999 984	0.999 985	0.999 985	0.999 986
4.20	0.999 987	0.999 987	0.999 988	0.999 988	0.999 989	0.999 989	0.999 990	0.999 990	0.999 991	0.999 991

续表

x	0.00	0.01	0.02	0.03	0.04	0.05	0.06	0.07	0.08	0.09
4.30	0.999 991	0.999 992	0.999 992	0.999 993	0.999 993	0.999 993	0.999 993	0.999 994	0.999 994	0.999 994
4.40	0.999 995	0.999 995	0.999 995	0.999 995	0.999 996	0.999 996	0.999 996	0.999 996	0.999 996	0.999 996
4.50	0.999 997	0.999 997	0.999 997	0.999 997	0.999 997	0.999 997	0.999 997	0.999 998	0.999 998	0.999 998
4.60	0.999 998	0.999 998	0.999 998	0.999 998	0.999 998	0.999 998	0.999 998	0.999 998	0.999 999	0.999 999
4.70	0.999 999	0.999 999	0.999 999	0.999 999	0.999 999	0.999 999	0.999 999	0.999 999	0.999 999	0.999 999
4.80	0.999 999	0.999 999	0.999 999	0.999 999	0.999 999	0.999 999	0.999 999	0.999 999	0.999 999	0.999 999
4.90	1.000 000	1.000 000	1.000 000	1.000 000	1.000 000	1.000 000	1.000 000	1.000 000	1.000 000	1.000 000

Note：This table gives $\Phi(x)$ for every x. For example：when $x=1.33$, $\Phi(x)=0.908\,241$.

Table 2 Table of Poisson Distribution Function

$$P(X \leqslant m) = \sum_{k=0}^{m} \frac{\lambda^k}{k!} e^{-\lambda}$$

m \ λ	0.1	0.2	0.3	0.4	0.5	0.6	0.7	0.8
0	0.904 837	0.818 731	0.740 818	0.670 320	0.606 531	0.548 812	0.496 585	0.449 329
1	0.090 484	0.163 746	0.222 245	0.268 128	0.303 265	0.329 287	0.347 610	0.359 463
2	0.004 524	0.016 375	0.033 337	0.053 626	0.075 816	0.098 786	0.121 663	0.143 785
3	0.000 151	0.001 092	0.003 334	0.007 150	0.012 636	0.019 757	0.028 388	0.038 343
4	0.000 004	0.000 055	0.000 250	0.000 715	0.001 580	0.002 964	0.004 968	0.007 669
5		0.000 002	0.000 015	0.000 057	0.000 158	0.000 356	0.000 696	0.001 227
6			0.000 001	0.000 004	0.000 013	0.000 036	0.000 081	0.000 164
7					0.000 001	0.000 003	0.000 008	0.000 019
8							0.000 001	0.000 002
9								

续表

m \ λ	0.9	1.0	1.5	2.0	2.5	3.0	3.5	4.0
0	0.406 570	0.367 879	0.223 130	0.135 335	0.082 085	0.049 787	0.030 197	0.018 316
1	0.365 913	0.367 879	0.334 695	0.270 671	0.205 212	0.149 361	0.105 691	0.073 263
2	0.164 661	0.183 940	0.251 021	0.270 671	0.256 516	0.224 042	0.184 959	0.146 525
3	0.049 398	0.061 313	0.125 511	0.180 447	0.213 763	0.224 042	0.215 785	0.195 367
4	0.011 115	0.015 328	0.047 067	0.090 224	0.133 602	0.168 031	0.188 812	0.195 367
5	0.002 001	0.003 066	0.014 120	0.036 089	0.066 801	0.100 819	0.132 169	0.156 293
6	0.000 300	0.000 511	0.003 530	0.012 030	0.027 834	0.050 409	0.077 098	0.104 196
7	0.000 039	0.000 073	0.000 756	0.003 437	0.009 941	0.021 604	0.038 549	0.059 540
8	0.000 004	0.000 009	0.000 142	0.000 859	0.003 106	0.008 102	0.016 865	0.029 770
9		0.000 001	0.000 024	0.000 191	0.000 863	0.002 701	0.006 559	0.013 231
10			0.000 004	0.000 038	0.000 216	0.000 810	0.002 296	0.005 292
11				0.000 007	0.000 049	0.000 221	0.000 730	0.001 925
12				0.000 001	0.000 010	0.000 055	0.000 213	0.000 642
13					0.000 002	0.000 013	0.000 057	0.000 197
14						0.000 003	0.000 014	0.000 056
15						0.000 001	0.000 003	0.000 015
16							0.000 001	0.000 004
17								0.000 001

续 表

m \ λ	4.5	5.0	5.5	6.0	6.5	7.0	7.5	8.0
0	0.011 109	0.006 738	0.004 087	0.002 479	0.001 503	0.000 912	0.000 553	0.000 335
1	0.049 990	0.033 690	0.022 477	0.014 873	0.009 772	0.006 383	0.004 148	0.002 684
2	0.112 479	0.084 224	0.061 812	0.044 618	0.031 760	0.022 341	0.015 555	0.010 735
3	0.168 718	0.140 374	0.113 323	0.089 235	0.068 814	0.052 129	0.038 889	0.028 626
4	0.189 808	0.175 467	0.155 819	0.133 853	0.111 822	0.091 226	0.072 916	0.057 252
5	0.170 827	0.175 467	0.171 401	0.160 623	0.145 369	0.127 717	0.109 375	0.091 604
6	0.128 120	0.146 223	0.157 117	0.160 623	0.157 483	0.149 003	0.136 718	0.122 138
7	0.082 363	0.104 445	0.123 449	0.137 677	0.146 234	0.149 003	0.146 484	0.139 587
8	0.046 329	0.065 278	0.084 871	0.103 258	0.118 815	0.130 377	0.137 329	0.139 587
9	0.023 165	0.036 266	0.051 866	0.068 838	0.085 811	0.101 405	0.114 440	0.124 077
10	0.010 424	0.018 133	0.028 526	0.041 303	0.055 777	0.070 983	0.085 830	0.099 262
11	0.004 264	0.008 242	0.014 263	0.022 529	0.032 959	0.045 171	0.058 521	0.072 190
12	0.001 599	0.003 434	0.006 537	0.011 264	0.017 853	0.026 350	0.036 575	0.048 127
13	0.000 554	0.001 321	0.002 766	0.005 199	0.008 926	0.014 188	0.021 101	0.029 616
14	0.000 178	0.000 472	0.001 087	0.002 228	0.004 144	0.007 094	0.011 304	0.016 924
15	0.000 053	0.000 157	0.000 398	0.000 891	0.001 796	0.003 311	0.005 652	0.009 026
16	0.000 015	0.000 049	0.000 137	0.000 334	0.000 730	0.001 448	0.002 649	0.004 513
17	0.000 004	0.000 014	0.000 044	0.000 118	0.000 279	0.000 596	0.001 169	0.002 124
18	0.000 001	0.000 004	0.000 014	0.000 039	0.000 101	0.000 232	0.000 487	0.000 944
19		0.000 001	0.000 004	0.000 012	0.000 034	0.000 085	0.000 192	0.000 397
20			0.000 001	0.000 004	0.000 011	0.000 030	0.000 072	0.000 159
21				0.000 001	0.000 003	0.000 010	0.000 026	0.000 061
22					0.000 001	0.000 003	0.000 009	0.000 022
23						0.000 001	0.000 003	0.000 008
24							0.000 001	0.000 003
25								0.000 001

续表

m \ λ	8.5	9.0	9.5	10	12	15	18	20
0	0.000 203	0.000 123	0.000 075	0.000 045	0.000 006	0.000 000	0.000 000	0.000 000
1	0.001 729	0.001 111	0.000 711	0.000 454	0.000 074	0.000 005	0.000 000	0.000 000
2	0.007 350	0.004 998	0.003 378	0.002 270	0.000 442	0.000 034	0.000 002	0.000 000
3	0.020 826	0.014 994	0.010 696	0.007 567	0.001 770	0.000 172	0.000 015	0.000 003
4	0.044 255	0.033 737	0.025 403	0.018 917	0.005 309	0.000 645	0.000 067	0.000 014
5	0.075 233	0.060 727	0.048 266	0.037 833	0.012 741	0.001 936	0.000 240	0.000 055
6	0.106 581	0.091 090	0.076 421	0.063 055	0.025 481	0.004 839	0.000 719	0.000 183
7	0.129 419	0.117 116	0.103 714	0.090 079	0.043 682	0.010 370	0.001 850	0.000 523
8	0.137 508	0.131 756	0.123 160	0.112 599	0.065 523	0.019 444	0.004 163	0.001 309
9	0.129 869	0.131 756	0.130 003	0.125 110	0.087 364	0.032 407	0.008 325	0.002 908
10	0.110 388	0.118 580	0.123 502	0.125 110	0.104 837	0.048 611	0.014 985	0.005 816
11	0.085 300	0.097 020	0.106 661	0.113 736	0.114 368	0.066 287	0.024 521	0.010 575
12	0.060 421	0.072 765	0.084 440	0.094 780	0.114 368	0.082 859	0.036 782	0.017 625
13	0.039 506	0.050 376	0.061 706	0.072 908	0.105 570	0.095 607	0.050 929	0.027 116
14	0.023 986	0.032 384	0.041 872	0.052 077	0.090 489	0.102 436	0.065 480	0.038 737
15	0.013 592	0.019 431	0.026 519	0.034 718	0.072 391	0.102 436	0.078 576	0.051 649
16	0.007 221	0.010 930	0.015 746	0.021 699	0.054 293	0.096 034	0.088 397	0.064 561
17	0.003 610	0.005 786	0.008 799	0.012 764	0.038 325	0.084 736	0.093 597	0.075 954
18	0.001 705	0.002 893	0.004 644	0.007 091	0.025 550	0.070 613	0.093 597	0.084 394
19	0.000 763	0.001 370	0.002 322	0.003 732	0.016 137	0.055 747	0.088 671	0.088 835

续表

m \ λ	8.5	9.0	9.5	10	12	15	18	20
20	0.000 324	0.000 617	0.001 103	0.001 866	0.009 682	0.041 810	0.079 804	0.088 835
21	0.000 131	0.000 264	0.000 499	0.000 889	0.005 533	0.029 865	0.068 403	0.084 605
22	0.000 051	0.000 108	0.000 215	0.000 404	0.003 018	0.020 362	0.055 966	0.076 914
23	0.000 019	0.000 042	0.000 089	0.000 176	0.001 574	0.013 280	0.043 800	0.066 881
24	0.000 007	0.000 016	0.000 035	0.000 073	0.000 787	0.008 300	0.032 850	0.055 735
25	0.000 002	0.000 006	0.000 013	0.000 029	0.000 378	0.004 980	0.023 652	0.044 588
26	0.000 001	0.000 002	0.000 005	0.000 011	0.000 174	0.002 873	0.016 374	0.034 298
27		0.000 001	0.000 002	0.000 004	0.000 078	0.001 596	0.010 916	0.025 406
28			0.000 001	0.000 001	0.000 033	0.000 855	0.007 018	0.018 147
29				0.000 001	0.000 014	0.000 442	0.004 356	0.012 515
30					0.000 005	0.000 221	0.002 613	0.008 344
31					0.000 002	0.000 107	0.001 517	0.005 383
32					0.000 001	0.000 050	0.000 854	0.003 364
33						0.000 023	0.000 466	0.002 039
34						0.000 010	0.000 246	0.001 199
35						0.000 004	0.000 127	0.000 685
36						0.000 002	0.000 063	0.000 381
37						0.000 001	0.000 031	0.000 206
38							0.000 015	0.000 108
39							0.000 007	0.000 056

Table 3　Table of the t Distribution

$P\{t(n)>t_\alpha(n)\}=\alpha$

n	$\alpha=0.10$	0.05	0.025	0.01	0.005	n	$\alpha=0.10$	0.05	0.025	0.01	0.005
1	3.0777	6.3138	12.7062	31.8205	63.6567	24	1.3178	1.7109	2.0639	2.4922	2.7969
2	1.8856	2.9200	4.3027	6.9646	9.9248	25	1.3163	1.7081	2.0595	2.4851	2.7874
3	1.6377	2.3534	3.1824	4.5407	5.8409						
4	1.5332	2.1318	2.7764	3.7469	4.6041	26	1.3150	1.7056	2.0555	2.4786	2.7787
5	1.4759	2.0150	2.5706	3.3649	4.0321	27	1.3137	1.7033	2.0518	2.4727	2.7707
						28	1.3125	1.7011	2.0484	2.4671	2.7633
6	1.4398	1.9432	2.4469	3.1427	3.7074	29	1.3114	1.6991	2.0452	2.4620	2.7564
7	1.4149	1.8946	2.3646	2.9980	3.4995	30	1.3104	1.6973	2.0423	2.4573	2.7500
8	1.3968	1.8595	2.3060	2.8965	3.3554						
9	1.3830	1.8331	2.2622	2.8214	3.2498	31	1.3095	1.6955	2.0395	2.4528	2.7440
10	1.3722	1.8125	2.2281	2.7638	3.1693	32	1.3086	1.6939	2.0369	2.4487	2.7385

续表

n	$\alpha=0.10$	0.05	0.025	0.01	0.005	n	$\alpha=0.10$	0.05	0.025	0.01	0.005
						33	1.3077	1.6924	2.0345	2.4448	2.7333
11	1.3634	1.7959	2.2010	2.7181	3.1058	34	1.3070	1.6909	2.0322	2.4411	2.7284
12	1.3562	1.7823	2.1788	2.6810	3.0545	35	1.3062	1.6896	2.0301	2.4377	2.7238
13	1.3502	1.7709	2.1604	2.6503	3.0123						
14	1.3450	1.7613	2.1448	2.6245	2.9768	36	1.3055	1.6883	2.0281	2.4345	2.7195
15	1.3406	1.7531	2.1314	2.6025	2.9467	37	1.3049	1.6871	2.0262	2.4314	2.7154
						38	1.3042	1.6860	2.0244	2.4286	2.7116
16	1.3368	1.7459	2.1199	2.5835	2.9208	39	1.3036	1.6849	2.0227	2.4258	2.7079
17	1.3334	1.7396	2.1098	2.5669	2.8982	40	1.3031	1.6839	2.0211	2.4233	2.7045
18	1.3304	1.7341	2.1009	2.5524	2.8784						
19	1.3277	1.7291	2.0930	2.5395	2.8609	41	1.3025	1.6829	2.0195	2.4208	2.7012
20	1.3253	1.7247	2.0860	2.5280	2.8453	42	1.3020	1.6820	2.0181	2.4185	2.6981
						43	1.3016	1.6811	2.0167	2.4163	2.6951
21	1.3232	1.7207	2.0796	2.5176	2.8314	44	1.3011	1.6802	2.0154	2.4141	2.6923
22	1.3212	1.7171	2.0739	2.5083	2.8188	45	1.3006	1.6794	2.0141	2.4121	2.6896
23	1.3195	1.7139	2.0687	2.4999	2.8073						

Table 4 Table of the χ^2 Distribution

$P\{\chi^2(n) > \chi^2_\alpha(n)\} = \alpha$

n	α=0.995	0.99	0.975	0.95	0.90	0.75	0.25	0.1	0.05	0.025	0.01	0.005
1	—	—	0.001	0.004	0.016	0.102	1.323	2.706	3.841	5.024	6.635	7.879
2	0.010	0.020	0.051	0.103	0.211	0.575	2.773	4.605	5.991	7.378	9.210	10.597
3	0.072	0.115	0.216	0.352	0.584	1.213	4.108	6.251	7.815	9.348	11.345	12.838
4	0.207	0.297	0.484	0.711	1.064	1.923	5.385	7.779	9.488	11.143	13.277	14.860
5	0.412	0.554	0.831	1.145	1.610	2.675	6.626	9.236	11.070	12.833	15.086	16.750
6	0.676	0.872	1.237	1.635	2.204	3.455	7.841	10.645	12.592	14.449	16.812	18.548
7	0.989	1.239	1.690	2.167	2.833	4.255	9.037	12.017	14.067	16.013	18.475	20.278
8	1.344	1.646	2.180	2.733	3.490	5.071	10.219	13.362	15.507	17.535	20.090	21.955
9	1.735	2.088	2.700	3.325	4.168	5.899	11.389	14.684	16.919	19.023	21.666	23.589
10	2.156	2.558	3.247	3.940	4.865	6.737	12.549	15.987	18.307	20.483	23.209	25.188

续表

n	α=0.995	0.99	0.975	0.95	0.90	0.75	0.25	0.1	0.05	0.025	0.01	0.005
11	2.603	3.053	3.816	4.575	5.578	7.584	13.701	17.275	19.675	21.920	24.725	26.757
12	3.074	3.571	4.404	5.226	6.304	8.438	14.845	18.549	21.026	23.337	26.217	28.300
13	3.565	4.107	5.009	5.892	7.042	9.299	15.984	19.812	22.362	24.736	27.688	29.819
14	4.075	4.660	5.629	6.571	7.790	10.165	17.117	21.064	23.685	26.119	29.141	31.319
15	4.601	5.229	6.262	7.261	8.547	11.037	18.245	22.307	24.996	27.488	30.578	32.801
16	5.142	5.812	6.908	7.962	9.312	11.912	19.369	23.542	26.296	28.845	32.000	34.267
17	5.697	6.408	7.564	8.672	10.085	12.792	20.489	24.769	27.587	30.191	33.409	35.718
18	6.265	7.015	8.231	9.390	10.865	13.675	21.605	25.989	28.869	31.526	34.805	37.156
19	6.844	7.633	8.907	10.117	11.651	14.562	22.718	27.204	30.144	32.852	36.191	38.582
20	7.434	8.260	9.591	10.851	12.443	15.452	23.828	28.412	31.410	34.170	37.566	39.997
21	8.034	8.897	10.283	11.591	13.240	16.344	24.935	29.615	32.671	35.479	38.932	41.401
22	8.643	9.542	10.982	12.338	14.042	17.240	26.039	30.813	33.924	36.781	40.289	42.796
23	9.260	10.196	11.689	13.091	14.848	18.137	27.141	32.007	35.172	38.076	41.638	44.181
24	9.886	10.856	12.401	13.848	15.659	19.037	28.241	33.196	36.415	39.364	42.980	45.559
25	10.520	11.524	13.120	14.611	16.473	19.939	29.339	34.382	37.652	40.646	44.314	46.928
26	11.160	12.198	13.844	15.379	17.292	20.843	30.435	35.563	38.885	41.923	45.642	48.290
27	11.808	12.879	14.573	16.151	18.114	21.749	31.528	36.741	40.113	43.195	46.963	49.645

续表

n	α=0.995	0.99	0.975	0.95	0.90	0.75	0.25	0.1	0.05	0.025	0.01	0.005
28	12.461	13.565	15.308	16.928	18.939	22.657	32.620	37.916	41.337	44.461	48.278	50.993
29	13.121	14.256	16.047	17.708	19.768	23.567	33.711	39.087	42.557	45.722	49.588	52.336
30	13.787	14.953	16.791	18.493	20.599	24.478	34.800	40.256	43.773	46.979	50.892	53.672
31	14.458	15.655	17.539	19.281	21.434	25.390	35.887	41.422	44.985	48.232	52.191	55.003
32	15.134	16.362	18.291	20.072	22.271	26.304	36.973	42.585	46.194	49.480	53.486	56.328
33	15.815	17.074	19.047	20.867	23.110	27.219	38.058	43.745	47.400	50.725	54.776	57.648
34	16.501	17.789	19.806	21.664	23.952	28.136	39.141	44.903	48.602	51.966	56.061	58.964
35	17.192	18.509	20.569	22.465	24.797	29.054	40.223	46.059	49.802	53.203	57.342	60.275
36	17.887	19.233	21.336	23.269	25.643	29.973	41.304	47.212	50.998	54.437	58.619	61.581
37	18.586	19.960	22.106	24.075	26.492	30.893	42.383	48.363	52.192	55.668	59.893	62.883
38	19.289	20.691	22.878	24.884	27.343	31.815	43.462	49.513	53.384	56.896	61.162	64.181
39	19.996	21.426	23.654	25.695	28.196	32.737	44.539	50.660	54.572	58.120	62.428	65.476
40	20.707	22.164	24.433	26.509	29.051	33.660	45.616	51.805	55.758	59.342	63.691	66.766
41	21.421	22.906	25.215	27.326	29.907	34.585	46.692	52.949	56.942	60.561	64.950	68.053
42	22.138	23.650	25.999	28.144	30.765	35.510	47.766	54.090	58.124	61.777	66.206	69.336
43	22.859	24.398	26.785	28.965	31.625	36.436	48.840	55.230	59.304	62.990	67.459	70.616
44	23.584	25.148	27.575	29.787	32.487	37.363	49.913	56.369	60.481	64.201	68.710	71.893
45	24.311	25.901	28.366	30.612	33.350	38.291	50.985	57.505	61.656	65.410	69.957	73.166

Table 5 Table of the F Distribution

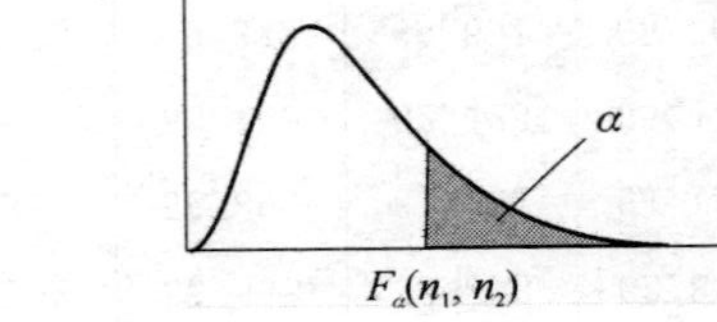

$$P\{F(n_1, n_2) > F_\alpha(n_1, n_2)\} = \alpha$$

(α=0.10)

n_2 \ n_1	1	2	3	4	5	6	7	8	9	10	12	15	20	24	30	40	60	120	∞
1	39.86	49.50	53.59	55.83	57.24	58.20	58.91	59.44	59.86	60.19	60.71	61.22	61.74	62.00	62.26	62.53	62.79	63.06	63.33
2	8.53	9.00	9.16	9.24	9.29	9.33	9.35	9.37	9.38	9.39	9.41	9.42	9.44	9.45	9.46	9.47	9.47	9.48	9.49
3	5.54	5.46	5.39	5.34	5.31	5.28	5.27	5.25	5.24	5.23	5.22	5.20	5.18	5.18	5.17	5.16	5.15	5.14	5.13
4	4.54	4.32	4.19	4.11	4.05	4.01	3.98	3.95	3.94	3.92	3.90	3.87	3.84	3.83	3.82	3.80	3.79	3.78	3.76
5	4.06	3.78	3.62	3.52	3.45	3.40	3.37	3.34	3.32	3.30	3.27	3.24	3.21	3.19	3.17	3.16	3.14	3.12	3.10
6	3.78	3.46	3.29	3.18	3.11	3.05	3.01	2.98	2.96	2.94	2.90	2.87	2.84	2.82	2.80	2.78	2.76	2.74	2.72
7	3.59	3.26	3.07	2.96	2.88	2.83	2.78	2.75	2.72	2.70	2.67	2.63	2.59	2.58	2.56	2.54	2.51	2.49	2.47
8	3.46	3.11	2.92	2.81	2.73	2.67	2.62	2.59	2.56	2.54	2.50	2.46	2.42	2.40	2.38	2.36	2.34	2.32	2.29
9	3.36	3.01	2.81	2.69	2.61	2.55	2.51	2.47	2.44	2.42	2.38	2.34	2.30	2.28	2.25	2.23	2.21	2.18	2.16
10	3.29	2.92	2.73	2.61	2.52	2.46	2.41	2.38	2.35	2.32	2.28	2.24	2.20	2.18	2.16	2.13	2.11	2.08	2.06

续表

n_2 \ n_1	1	2	3	4	5	6	7	8	9	10	12	15	20	24	30	40	60	120	∞
11	3.23	2.86	2.66	2.54	2.45	2.39	2.34	2.30	2.27	2.25	2.21	2.17	2.12	2.10	2.08	2.05	2.03	2.00	1.97
12	3.18	2.81	2.61	2.48	2.39	2.33	2.28	2.24	2.21	2.19	2.15	2.10	2.06	2.04	2.01	1.99	1.96	1.93	1.90
13	3.14	2.76	2.56	2.43	2.35	2.28	2.23	2.20	2.16	2.14	2.10	2.05	2.01	1.98	1.96	1.93	1.90	1.88	1.85
14	3.10	2.73	2.52	2.39	2.31	2.24	2.19	2.15	2.12	2.10	2.05	2.01	1.96	1.94	1.91	1.89	1.86	1.83	1.80
15	3.07	2.70	2.49	2.36	2.27	2.21	2.16	2.12	2.09	2.06	2.02	1.97	1.92	1.90	1.87	1.85	1.82	1.79	1.76
16	3.05	2.67	2.46	2.33	2.24	2.18	2.13	2.09	2.06	2.03	1.99	1.94	1.89	1.87	1.84	1.81	1.78	1.75	1.72
17	3.03	2.64	2.44	2.31	2.22	2.15	2.10	2.06	2.03	2.00	1.96	1.91	1.86	1.84	1.81	1.78	1.75	1.72	1.69
18	3.01	2.62	2.42	2.29	2.20	2.13	2.08	2.04	2.00	1.98	1.93	1.89	1.84	1.81	1.78	1.75	1.72	1.69	1.66
19	2.99	2.61	2.40	2.27	2.18	2.11	2.06	2.02	1.98	1.96	1.91	1.86	1.81	1.79	1.76	1.73	1.70	1.67	1.63
20	2.97	2.59	2.38	2.25	2.16	2.09	2.04	2.00	1.96	1.94	1.89	1.84	1.79	1.77	1.74	1.71	1.68	1.64	1.61
21	2.96	2.57	2.36	2.23	2.14	2.08	2.02	1.98	1.95	1.92	1.87	1.83	1.78	1.75	1.72	1.69	1.66	1.62	1.59
22	2.95	2.56	2.35	2.22	2.13	2.06	2.01	1.97	1.93	1.90	1.86	1.81	1.76	1.73	1.70	1.67	1.64	1.60	1.57
23	2.94	2.55	2.34	2.21	2.11	2.05	1.99	1.95	1.92	1.89	1.84	1.80	1.74	1.72	1.69	1.66	1.62	1.59	1.55
24	2.93	2.54	2.33	2.19	2.10	2.04	1.98	1.94	1.91	1.88	1.83	1.78	1.73	1.70	1.67	1.64	1.61	1.57	1.53
25	2.92	2.53	2.32	2.18	2.09	2.02	1.97	1.93	1.89	1.87	1.82	1.77	1.72	1.69	1.66	1.63	1.59	1.56	1.52
26	2.91	2.52	2.31	2.17	2.08	2.01	1.96	1.92	1.88	1.86	1.81	1.76	1.71	1.68	1.65	1.61	1.58	1.54	1.50
27	2.90	2.51	2.30	2.17	2.07	2.00	1.95	1.91	1.87	1.85	1.80	1.75	1.70	1.67	1.64	1.60	1.57	1.53	1.49
28	2.89	2.50	2.29	2.16	2.06	2.00	1.94	1.90	1.87	1.84	1.79	1.74	1.69	1.66	1.63	1.59	1.56	1.52	1.48
29	2.89	2.50	2.28	2.15	2.06	1.99	1.93	1.89	1.86	1.83	1.78	1.73	1.68	1.65	1.62	1.58	1.55	1.51	1.47

续表

n_2 \ n_1	1	2	3	4	5	6	7	8	9	10	12	15	20	24	30	40	60	120	∞
30	2.88	2.49	2.28	2.14	2.05	1.98	1.93	1.88	1.85	1.82	1.77	1.72	1.67	1.64	1.61	1.57	1.54	1.50	1.46
40	2.84	2.44	2.23	2.09	2.00	1.93	1.87	1.83	1.79	1.76	1.71	1.66	1.61	1.57	1.54	1.51	1.47	1.42	1.38
60	2.79	2.39	2.18	2.04	1.95	1.87	1.82	1.77	1.74	1.71	1.66	1.60	1.54	1.51	1.48	1.44	1.40	1.35	1.29
120	2.75	2.35	2.13	1.99	1.90	1.82	1.77	1.72	1.68	1.65	1.60	1.55	1.48	1.45	1.41	1.37	1.32	1.26	1.19
∞	2.71	2.30	2.08	1.94	1.85	1.77	1.72	1.67	1.63	1.60	1.55	1.49	1.42	1.38	1.34	1.30	1.24	1.17	1.00
(α=0.05)																			
1	161.45	199.50	215.71	224.58	230.16	233.99	236.77	238.88	240.54	241.88	243.91	245.95	248.01	249.05	250.10	251.14	252.20	253.25	254.31
2	18.51	19.00	19.16	19.25	19.30	19.33	19.35	19.37	19.38	19.40	19.41	19.43	19.45	19.45	19.46	19.47	19.48	19.49	19.50
3	10.13	9.55	9.28	9.12	9.01	8.94	8.89	8.85	8.81	8.79	8.74	8.70	8.66	8.64	8.62	8.59	8.57	8.55	8.53
4	7.71	6.94	6.59	6.39	6.26	6.16	6.09	6.04	6.00	5.96	5.91	5.86	5.80	5.77	5.75	5.72	5.69	5.66	5.63
5	6.61	5.79	5.41	5.19	5.05	4.95	4.88	4.82	4.77	4.74	4.68	4.62	4.56	4.53	4.50	4.46	4.43	4.40	4.36
6	5.99	5.14	4.76	4.53	4.39	4.28	4.21	4.15	4.10	4.06	4.00	3.94	3.87	3.84	3.81	3.77	3.74	3.70	3.67
7	5.59	4.74	4.35	4.12	3.97	3.87	3.79	3.73	3.68	3.64	3.57	3.51	3.44	3.41	3.38	3.34	3.30	3.27	3.23
8	5.32	4.46	4.07	3.84	3.69	3.58	3.50	3.44	3.39	3.35	3.28	3.22	3.15	3.12	3.08	3.04	3.01	2.97	2.93
9	5.12	4.26	3.86	3.63	3.48	3.37	3.29	3.23	3.18	3.14	3.07	3.01	2.94	2.90	2.86	2.83	2.79	2.75	2.71
10	4.96	4.10	3.71	3.48	3.33	3.22	3.14	3.07	3.02	2.98	2.91	2.85	2.77	2.74	2.70	2.66	2.62	2.58	2.54
11	4.84	3.98	3.59	3.36	3.20	3.09	3.01	2.95	2.90	2.85	2.79	2.72	2.65	2.61	2.57	2.53	2.49	2.45	2.40
12	4.75	3.89	3.49	3.26	3.11	3.00	2.91	2.85	2.80	2.75	2.69	2.62	2.54	2.51	2.47	2.43	2.38	2.34	2.30
13	4.67	3.81	3.41	3.18	3.03	2.92	2.83	2.77	2.71	2.67	2.60	2.53	2.46	2.42	2.38	2.34	2.30	2.25	2.21

续表

n_2 \ n_1	1	2	3	4	5	6	7	8	9	10	12	15	20	24	30	40	60	120	∞
14	4.60	3.74	3.34	3.11	2.96	2.85	2.76	2.70	2.65	2.60	2.53	2.46	2.39	2.35	2.31	2.27	2.22	2.18	2.13
15	4.54	3.68	3.29	3.06	2.90	2.79	2.71	2.64	2.59	2.54	2.48	2.40	2.33	2.29	2.25	2.20	2.16	2.11	2.07
16	4.49	3.63	3.24	3.01	2.85	2.74	2.66	2.59	2.54	2.49	2.42	2.35	2.28	2.24	2.19	2.15	2.11	2.06	2.01
17	4.45	3.59	3.20	2.96	2.81	2.70	2.61	2.55	2.49	2.45	2.38	2.31	2.23	2.19	2.15	2.10	2.06	2.01	1.96
18	4.41	3.55	3.16	2.93	2.77	2.66	2.58	2.51	2.46	2.41	2.34	2.27	2.19	2.15	2.11	2.06	2.02	1.97	1.92
19	4.38	3.52	3.13	2.90	2.74	2.63	2.54	2.48	2.42	2.38	2.31	2.23	2.16	2.11	2.07	2.03	1.98	1.93	1.88
20	4.35	3.49	3.10	2.87	2.71	2.60	2.51	2.45	2.39	2.35	2.28	2.20	2.12	2.08	2.04	1.99	1.95	1.90	1.84
21	4.32	3.47	3.07	2.84	2.68	2.57	2.49	2.42	2.37	2.32	2.25	2.18	2.10	2.05	2.01	1.96	1.92	1.87	1.81
22	4.30	3.44	3.05	2.82	2.66	2.55	2.46	2.40	2.34	2.30	2.23	2.15	2.07	2.03	1.98	1.94	1.89	1.84	1.78
23	4.28	3.42	3.03	2.80	2.64	2.53	2.44	2.37	2.32	2.27	2.20	2.13	2.05	2.01	1.96	1.91	1.86	1.81	1.76
24	4.26	3.40	3.01	2.78	2.62	2.51	2.42	2.36	2.30	2.25	2.18	2.11	2.03	1.98	1.94	1.89	1.84	1.79	1.73
25	4.24	3.39	2.99	2.76	2.60	2.49	2.40	2.34	2.28	2.24	2.16	2.09	2.01	1.96	1.92	1.87	1.82	1.77	1.71
26	4.23	3.37	2.98	2.74	2.59	2.47	2.39	2.32	2.27	2.22	2.15	2.07	1.99	1.95	1.90	1.85	1.80	1.75	1.69
27	4.21	3.35	2.96	2.73	2.57	2.46	2.37	2.31	2.25	2.20	2.13	2.06	1.97	1.93	1.88	1.84	1.79	1.73	1.67
28	4.20	3.34	2.95	2.71	2.56	2.45	2.36	2.29	2.24	2.19	2.12	2.04	1.96	1.91	1.87	1.82	1.77	1.71	1.65
29	4.18	3.33	2.93	2.70	2.55	2.43	2.35	2.28	2.22	2.18	2.10	2.03	1.94	1.90	1.85	1.81	1.75	1.70	1.64
30	4.17	3.32	2.92	2.69	2.53	2.42	2.33	2.27	2.21	2.16	2.09	2.01	1.93	1.89	1.84	1.79	1.74	1.68	1.62
40	4.08	3.23	2.84	2.61	2.45	2.34	2.25	2.18	2.12	2.08	2.00	1.92	1.84	1.79	1.74	1.69	1.64	1.58	1.51
60	4.00	3.15	2.76	2.53	2.37	2.25	2.17	2.10	2.04	1.99	1.92	1.84	1.75	1.70	1.65	1.59	1.53	1.47	1.39
120	3.92	3.07	2.68	2.45	2.29	2.18	2.09	2.02	1.96	1.91	1.83	1.75	1.66	1.61	1.55	1.50	1.43	1.35	1.25
∞	3.84	3.00	2.60	2.37	2.21	2.10	2.01	1.94	1.88	1.83	1.75	1.67	1.57	1.52	1.46	1.39	1.32	1.22	1.00

续表

(α=0.025)																			
1	647.79	799.50	864.16	899.58	921.85	937.11	948.22	956.66	963.28	968.63	976.71	984.87	993.10	997.25	1001.41	1005.60	1009.80	1014.02	1018.26
2	38.51	39.00	39.17	39.25	39.30	39.33	39.36	39.37	39.39	39.40	39.41	39.43	39.45	39.46	39.46	39.47	39.48	39.49	39.50
3	17.44	16.04	15.44	15.10	14.88	14.73	14.62	14.54	14.47	14.42	14.34	14.25	14.17	14.12	14.08	14.04	13.99	13.95	13.90
4	12.22	10.65	9.98	9.60	9.36	9.20	9.07	8.98	8.90	8.84	8.75	8.66	8.56	8.51	8.46	8.41	8.36	8.31	8.26
5	10.01	8.43	7.76	7.39	7.15	6.98	6.85	6.76	6.68	6.62	6.52	6.43	6.33	6.28	6.23	6.18	6.12	6.07	6.02
6	8.81	7.26	6.60	6.23	5.99	5.82	5.70	5.60	5.52	5.46	5.37	5.27	5.17	5.12	5.07	5.01	4.96	4.90	4.85
7	8.07	6.54	5.89	5.52	5.29	5.12	4.99	4.90	4.82	4.76	4.67	4.57	4.47	4.41	4.36	4.31	4.25	4.20	4.14
8	7.57	6.06	5.42	5.05	4.82	4.65	4.53	4.43	4.36	4.30	4.20	4.10	4.00	3.95	3.89	3.84	3.78	3.73	3.67
9	7.21	5.71	5.08	4.72	4.48	4.32	4.20	4.10	4.03	3.96	3.87	3.77	3.67	3.61	3.56	3.51	3.45	3.39	3.33
10	6.94	5.46	4.83	4.47	4.24	4.07	3.95	3.85	3.78	3.72	3.62	3.52	3.42	3.37	3.31	3.26	3.20	3.14	3.08
11	6.72	5.26	4.63	4.28	4.04	3.88	3.76	3.66	3.59	3.53	3.43	3.33	3.23	3.17	3.12	3.06	3.00	2.94	2.88
12	6.55	5.10	4.47	4.12	3.89	3.73	3.61	3.51	3.44	3.37	3.28	3.18	3.07	3.02	2.96	2.91	2.85	2.79	2.72
13	6.41	4.97	4.35	4.00	3.77	3.60	3.48	3.39	3.31	3.25	3.15	3.05	2.95	2.89	2.84	2.78	2.72	2.66	2.60
14	6.30	4.86	4.24	3.89	3.66	3.50	3.38	3.29	3.21	3.15	3.05	2.95	2.84	2.79	2.73	2.67	2.61	2.55	2.49
15	6.20	4.77	4.15	3.80	3.58	3.41	3.29	3.20	3.12	3.06	2.96	2.86	2.76	2.70	2.64	2.59	2.52	2.46	2.40
16	6.12	4.69	4.08	3.73	3.50	3.34	3.22	3.12	3.05	2.99	2.89	2.79	2.68	2.63	2.57	2.51	2.45	2.38	2.32
17	6.04	4.62	4.01	3.66	3.44	3.28	3.16	3.06	2.98	2.92	2.82	2.72	2.62	2.56	2.50	2.44	2.38	2.32	2.25
18	5.98	4.56	3.95	3.61	3.38	3.22	3.10	3.01	2.93	2.87	2.77	2.67	2.56	2.50	2.44	2.38	2.32	2.26	2.19
19	5.92	4.51	3.90	3.56	3.33	3.17	3.05	2.96	2.88	2.82	2.72	2.62	2.51	2.45	2.39	2.33	2.27	2.20	2.13
20	5.87	4.46	3.86	3.51	3.29	3.13	3.01	2.91	2.84	2.77	2.68	2.57	2.46	2.41	2.35	2.29	2.22	2.16	2.09

续表

21	5.83	4.42	3.82	3.48	3.25	3.09	2.97	2.87	2.80	2.73	2.64	2.53	2.42	2.37	2.31	2.25	2.18	2.11	2.04
22	5.79	4.38	3.78	3.44	3.22	3.05	2.93	2.84	2.76	2.70	2.60	2.50	2.39	2.33	2.27	2.21	2.14	2.08	2.00
23	5.75	4.35	3.75	3.41	3.18	3.02	2.90	2.81	2.73	2.67	2.57	2.47	2.36	2.30	2.24	2.18	2.11	2.04	1.97
24	5.72	4.32	3.72	3.38	3.15	2.99	2.87	2.78	2.70	2.64	2.54	2.44	2.33	2.27	2.21	2.15	2.08	2.01	1.94
25	5.69	4.29	3.69	3.35	3.13	2.97	2.85	2.75	2.68	2.61	2.51	2.41	2.30	2.24	2.18	2.12	2.05	1.98	1.91
26	5.66	4.27	3.67	3.33	3.10	2.94	2.82	2.73	2.65	2.59	2.49	2.39	2.28	2.22	2.16	2.09	2.03	1.95	1.88
27	5.63	4.24	3.65	3.31	3.08	2.92	2.80	2.71	2.63	2.57	2.47	2.36	2.25	2.19	2.13	2.07	2.00	1.93	1.85
28	5.61	4.22	3.63	3.29	3.06	2.90	2.78	2.69	2.61	2.55	2.45	2.34	2.23	2.17	2.11	2.05	1.98	1.91	1.83
29	5.59	4.20	3.61	3.27	3.04	2.88	2.76	2.67	2.59	2.53	2.43	2.32	2.21	2.15	2.09	2.03	1.96	1.89	1.81
30	5.57	4.18	3.59	3.25	3.03	2.87	2.75	2.65	2.57	2.51	2.41	2.31	2.20	2.14	2.07	2.01	1.94	1.87	1.79
40	5.42	4.05	3.46	3.13	2.90	2.74	2.62	2.53	2.45	2.39	2.29	2.18	2.07	2.01	1.94	1.88	1.80	1.72	1.64
60	5.29	3.93	3.34	3.01	2.79	2.63	2.51	2.41	2.33	2.27	2.17	2.06	1.94	1.88	1.82	1.74	1.67	1.58	1.48
120	5.15	3.80	3.23	2.89	2.67	2.52	2.39	2.30	2.22	2.16	2.05	1.94	1.82	1.76	1.69	1.61	1.53	1.43	1.31
∞	5.02	3.69	3.12	2.79	2.57	2.41	2.29	2.19	2.11	2.05	1.94	1.83	1.71	1.64	1.57	1.48	1.39	1.27	1.00

($\alpha=0.001$)

n_1 / n_2	1	2	3	4	5	6	7	8	9	10	12	15	20	24	30	40	60	120	∞
1	4 052	5 000	5 403	5 625	5 764	5 859	5 928	5 981	6 022	6 056	6 106	6 157	6 209	6 235	6 261	6 287	6 313	6 339	6 366
2	98.50	99.00	99.17	99.25	99.30	99.33	99.36	99.37	99.39	99.40	99.42	99.43	99.45	99.46	99.47	99.47	99.48	99.49	99.50
3	34.12	30.82	29.46	28.71	28.24	27.91	27.67	27.49	27.35	27.23	27.05	26.87	26.69	26.60	26.50	26.41	26.32	26.22	26.13
4	21.20	18.00	16.69	15.98	15.52	15.21	14.98	14.80	14.66	14.55	14.37	14.20	14.02	13.93	13.84	13.75	13.65	13.56	13.46
5	16.26	13.27	12.06	11.39	10.97	10.67	10.46	10.29	10.16	10.05	9.89	9.72	9.55	9.47	9.38	9.29	9.20	9.11	9.02

续表

n_2 \ n_1	1	2	3	4	5	6	7	8	9	10	12	15	20	24	30	40	60	120	∞
6	13.75	10.92	9.78	9.15	8.75	8.47	8.26	8.10	7.98	7.87	7.72	7.56	7.40	7.31	7.23	7.14	7.06	6.97	6.88
7	12.25	9.55	8.45	7.85	7.46	7.19	6.99	6.84	6.72	6.62	6.47	6.31	6.16	6.07	5.99	5.91	5.82	5.74	5.65
8	11.26	8.65	7.59	7.01	6.63	6.37	6.18	6.03	5.91	5.81	5.67	5.52	5.36	5.28	5.20	5.12	5.03	4.95	4.86
9	10.56	8.02	6.99	6.42	6.06	5.80	5.61	5.47	5.35	5.26	5.11	4.96	4.81	4.73	4.65	4.57	4.48	4.40	4.31
10	10.04	7.56	6.55	5.99	5.64	5.39	5.20	5.06	4.94	4.85	4.71	4.56	4.41	4.33	4.25	4.17	4.08	4.00	3.91
11	9.65	7.21	6.22	5.67	5.32	5.07	4.89	4.74	4.63	4.54	4.40	4.25	4.10	4.02	3.94	3.86	3.78	3.69	3.60
12	9.33	6.93	5.95	5.41	5.06	4.82	4.64	4.50	4.39	4.30	4.16	4.01	3.86	3.78	3.70	3.62	3.54	3.45	3.36
13	9.07	6.70	5.74	5.21	4.86	4.62	4.44	4.30	4.19	4.10	3.96	3.82	3.66	3.59	3.51	3.43	3.34	3.25	3.17
14	8.86	6.51	5.56	5.04	4.69	4.46	4.28	4.14	4.03	3.94	3.80	3.66	3.51	3.43	3.35	3.27	3.18	3.09	3.00
15	8.68	6.36	5.42	4.89	4.56	4.32	4.14	4.00	3.89	3.80	3.67	3.52	3.37	3.29	3.21	3.13	3.05	2.96	2.87
16	8.53	6.23	5.29	4.77	4.44	4.20	4.03	3.89	3.78	3.69	3.55	3.41	3.26	3.18	3.10	3.02	2.93	2.84	2.75
17	8.40	6.11	5.18	4.67	4.34	4.10	3.93	3.79	3.68	3.59	3.46	3.31	3.16	3.08	3.00	2.92	2.83	2.75	2.65
18	8.29	6.01	5.09	4.58	4.25	4.01	3.84	3.71	3.60	3.51	3.37	3.23	3.08	3.00	2.92	2.84	2.75	2.66	2.57
19	8.18	5.93	5.01	4.50	4.17	3.94	3.77	3.63	3.52	3.43	3.30	3.15	3.00	2.92	2.84	2.76	2.67	2.58	2.49
20	8.10	5.85	4.94	4.43	4.10	3.87	3.70	3.56	3.46	3.37	3.23	3.09	2.94	2.86	2.78	2.69	2.61	2.52	2.42
21	8.02	5.78	4.87	4.37	4.04	3.81	3.64	3.51	3.40	3.31	3.17	3.03	2.88	2.80	2.72	2.64	2.55	2.46	2.36
22	7.95	5.72	4.82	4.31	3.99	3.76	3.59	3.45	3.35	3.26	3.12	2.98	2.83	2.75	2.67	2.58	2.50	2.40	2.31
23	7.88	5.66	4.76	4.26	3.94	3.71	3.54	3.41	3.30	3.21	3.07	2.93	2.78	2.70	2.62	2.54	2.45	2.35	2.26
24	7.82	5.61	4.72	4.22	3.90	3.67	3.50	3.36	3.26	3.17	3.03	2.89	2.74	2.66	2.58	2.49	2.40	2.31	2.21
25	7.77	5.57	4.68	4.18	3.85	3.63	3.46	3.32	3.22	3.13	2.99	2.85	2.70	2.62	2.54	2.45	2.36	2.27	2.17

续表

n_2 \ n_1	1	2	3	4	5	6	7	8	9	10	12	15	20	24	30	40	60	120	∞
26	7.72	5.53	4.64	4.14	3.82	3.59	3.42	3.29	3.18	3.09	2.96	2.81	2.66	2.58	2.50	2.42	2.33	2.23	2.13
27	7.68	5.49	4.60	4.11	3.78	3.56	3.39	3.26	3.15	3.06	2.93	2.78	2.63	2.55	2.47	2.38	2.29	2.20	2.10
28	7.64	5.45	4.57	4.07	3.75	3.53	3.36	3.23	3.12	3.03	2.90	2.75	2.60	2.52	2.44	2.35	2.26	2.17	2.06
29	7.60	5.42	4.54	4.04	3.73	3.50	3.33	3.20	3.09	3.00	2.87	2.73	2.57	2.49	2.41	2.33	2.23	2.14	2.03
30	7.56	5.39	4.51	4.02	3.70	3.47	3.30	3.17	3.07	2.98	2.84	2.70	2.55	2.47	2.39	2.30	2.21	2.11	2.01
40	7.31	5.18	4.31	3.83	3.51	3.29	3.12	2.99	2.89	2.80	2.66	2.52	2.37	2.29	2.20	2.11	2.02	1.92	1.80
60	7.08	4.98	4.13	3.65	3.34	3.12	2.95	2.82	2.72	2.63	2.50	2.35	2.20	2.12	2.03	1.94	1.84	1.73	1.60
120	6.85	4.79	3.95	3.48	3.17	2.96	2.79	2.66	2.56	2.47	2.34	2.19	2.03	1.95	1.86	1.76	1.66	1.53	1.38
∞	6.63	4.61	3.78	3.32	3.02	2.80	2.64	2.51	2.41	2.32	2.18	2.04	1.88	1.79	1.70	1.59	1.47	1.32	1.00

Solutions for Exercises

Chapter 1

1.1. $\Omega=\{3,4,5,6,7,8,9,10,11,12,13,14,15,16,17,18\}$.

1.2. (1) AB^CC^C. (2) ABC^C. (3) $AB\cup AC\cup BC$.

(4) $A^CB^CC^C\cup AB^CC^C\cup A^CB^CC\cup A^CBC^C$.

1.3. (1) For any $\omega\in B^C$, if $\omega\in A$ then $\omega\in B$ since $A\subset B$. Then we have a contradiction that $\omega\in B$. Therefore, $\omega\in A^C$.

$B^C\subset A^C$ is immediate.

(2) $AB\cup AB^C=A(B\cup B^c)=A\Omega=A$.

(3) $(\bigcup\limits_{i=1}^{n-1} A_i\cup A_n)^C=(\bigcup\limits_{i=1}^{n-1} A_i)^C\cap A_n^C=\cdots=\bigcap\limits_{i=1}^{n} A_i^C$.

1.4. $S=\{r_1,r_2,\cdots,r_5,b_1,b_2,\cdots,b_5\}$.

$ABC=\{b_2\}$: a blue card numbered 2 is selected.

$A\cup B\cup C=\{b_1,b_2,b_3,b_4,b_5,r_1,r_2,r_4\}$: a blue ball, or red ball with even number or number 1, is selected.

$A(B\cup C)=\{r_2,b_2,b_4\}$: a blue ball with an even number, or a red ball with number 2 is selected.

$A^C(B\cup C)=\{b_1,b_3,b_5,r_1\}$: a blue ball with an odd number, or a red ball with number 1 is selected.

1.5. $A=E_1E_2^C$, $B=E_2E_1^C$, $O=E_1^CE_2^C$, $AB=E_1E_2$.

1.6. drawing lots: $\frac{a}{a+b}$.

1.7. A: at least two students have the same birthday.

A^c: all the students have different birthday.

$$P(A^C)=\frac{365!}{365^{100}(365-100)!}=\frac{365!}{365^{100}265!}.$$

$$P(A)=1-\frac{365!}{365^{100}265!}.$$

1.8. A: the students who are enrolled in a mathematics class.

B: the students who are enrolled in a English class.

C: the students who are enrolled in a music class.

$P(A)=7/15, P(B)=8/15, P(C)=9/15,$

$P(AB)=4/15, P(BC)=5/15, P(AC)=4/15, P(ABC)=2/15.$

$$P(A\cup B\cup C)=P(A)+P(B)+P(C)$$
$$-P(AB)-P(BC)-P(AC)+P(ABC)=13/15.$$

1.9. $P(A^CB)=P(B)-P(AB), \frac{P(A^CB)}{P(B)}=\frac{P(B)-P(AB)}{P(B)},$

$P(A^C|B)=1-P(A|B).$

1.10. $P(B|A^C)=\frac{P(A^CB)}{P(A^C)}=\frac{P(B)-P(AB)}{1-P(A)}=\frac{0.6-0.4}{06}=\frac{1}{3}.$

1.11. $P(A)=\frac{n}{m+n}\times\frac{n+k}{m+n+k}+\frac{m}{m+n}\times\frac{n}{n+m+k}.$

1.12. $P_n=C_{n-1}^4p^4q^{n-5}p=C_{n-1}^4p^5q^{n-5}.$

1.13. $P(A\cup B\cup C)=P(A)+P(B)+P(C)$

$$-P(AB)-P(BC)-P(AC)+P(ABC)=\frac{3}{4}.$$

1.14. A: this bolt is long. $P(A)=\frac{1}{2}\times\frac{40}{60}+\frac{1}{2}\times\frac{10}{30}=\frac{1}{2}.$

1.15. (1)A: the item is found to be defective.

$P(A)=10\%\times5\%+30\%\times3\%+60\%\times1\%=2\%.$

B: the item is produced by machine M_2.

$$P(B|A)=\frac{P(AB)}{P(A)}=\frac{30\%\times 3\%}{2\%}=45\%.$$

1.16. Yes. The probability of switching the choice is 2/3, not switching 1/2.

1.17. (1) The strategy is as below:

The gambler bets all his money, 100 dollars, for the first time. The probability of winning and leaving with 200 dollars is 1/2.

(2) The strategy is as below:

At step 1, the gambler bets all his money, 100 dollars, for the first time. The probability of winning 100 dollars is 1/2.

At step 2, he keeps 100 dollars and bets 100 dollars that he wins at the last step. Then after gambling this time, the probability that he has 300 dollars is 1/4. If this case occurs, he leaves. The probability that he has 100 dollars is 1/4. If this case occurs, repeat step 1.

Chapter 2

2.1. $P\{X\geqslant 1.5\}=0.3$.

2.2. $\lim\limits_{x\to\infty}F(x)=a\lim\limits_{x\to\infty}G(3x)+0.7\lim\limits_{x\to\infty}G(\frac{x}{2}+1)=a+0.7=1.\ a=0.3.$

2.3. (1) $P\{X \text{ is even}\}=\frac{1}{2}+\frac{1}{2^3}+\frac{1}{2^5}+\cdots=\lim\limits_{n\to\infty}\frac{\frac{1}{4}}{1-\frac{1}{4}}=\frac{1}{3}.$

(2) $P\{X\geqslant 5\}=\sum\limits_{k=5}^{\infty}\frac{1}{2^k}=\lim\limits_{k\to\infty}\frac{\frac{1}{2^5}}{1-\frac{1}{2}}=\frac{1}{16}.$

2.4. $p_1=P\{X \text{ is not even}\}, p_2=P\{X \text{ is even}\}, p_2+p_1=1.$

(1) $p_1=\sum\limits_{i=0}^{[n/2]}C_n^{2i+1}p^{2i+1}q^{n-(2i+1)},\ p_2=\sum\limits_{i=0}^{[n/2]}C_n^{2i}p^{2i}q^{n-2i}.\ q=1-p.$

$p_2-p_1=(-p+q)^n, p_2=((-p+q)^n+1)/2.$

(2) $p_2=\sum_{i=0}^{\infty}\frac{\lambda^{2i}}{(2i)!}\mathrm{e}^{-\lambda}, p_1=\sum_{i=0}^{\infty}\frac{\lambda^{2i+1}}{(2i+1)!}\mathrm{e}^{-\lambda}.$

$$p_2-p_1=\sum_{i=0}^{\infty}(-1)^i\frac{\lambda}{i!}^i\mathrm{e}^{-\lambda}=\mathrm{e}^{-\lambda}\mathrm{e}^{-\lambda}=\mathrm{e}^{-2\lambda}.$$

$P\{X \text{ is even}\}=\frac{1}{2}[1+\mathrm{e}^{-2\lambda}].$

2.5. $\mathrm{P}\{\mathrm{X}=0\}=\mathrm{q}^2=\frac{4}{9},\mathrm{q}=\frac{2}{3},\mathrm{p}=\frac{1}{3}.$

$$P\{Y\geqslant 1\}=\sum_{i=1}^{3}\mathrm{C}_3^i\frac{1}{3}^i\frac{2}{3}^{3-i}=\frac{19}{27}.$$

2.6. $P\{X\quad=i\}=\mathrm{C}_n^i p^i q^{n-i}=\frac{n!\ p^i q^{n-i}}{i!\ (n-i)!}$

$$\begin{cases}P\{X=i\}\geqslant P\{X=i+1\}\\P\{X=i\}\geqslant P\{X=i-1\}\end{cases}\Rightarrow\begin{cases}\frac{n!\ p^i q^{n-i}}{i!\ (n-i)!}\geqslant\frac{n!\ p^{i+1}q^{n-i-1}}{(i+1)!\ (n-i-1)!}\\\frac{n!\ p^i q^{n-i}}{i!\ (n-i)!}\geqslant\frac{n!\ p^{i-1}q^{n-i+1}}{(i-1)!\ (n-i+1)!}\end{cases}$$

$$\Rightarrow\begin{cases}i\geqslant(n+1)p-1\\i\leqslant(n+1)p\end{cases}\Rightarrow k=[(n+1)p].$$

2.7. $P\{X\quad=i\}=\frac{\lambda^i}{i!}\mathrm{e}^{-\lambda}.$

$$\begin{cases}P\{X=i\}\geqslant P\{X=i+1\}\\P\{X=i\}\geqslant P\{X=i-1\}\end{cases}\Rightarrow\begin{cases}\frac{\lambda^i}{i!}\mathrm{e}^{-\lambda}\geqslant\frac{\lambda^{i+1}}{(i+1)!}\mathrm{e}^{-\lambda}\\\frac{\lambda^i}{i!}\mathrm{e}^{-\lambda}\geqslant\frac{\lambda^{i-1}}{(i-1)!}\mathrm{e}^{-\lambda}\end{cases}\Rightarrow\lambda\geqslant i\geqslant\lambda-1.$$

$\Rightarrow k=[\lambda].$

2.8. $P\{X=1\}=P\{X=2\}\Rightarrow\frac{\lambda^1}{1!}\mathrm{e}^{-\lambda}=\frac{\lambda^2}{2!}\mathrm{e}^{-\lambda}\Rightarrow\lambda=2.$

$P\{Y\geqslant 1\}=1-P\{X=0\}=1-\mathrm{e}^{-2}.$

2.9. $P\{Y=2\}=P\{Y=2,X=2\}+P\{Y=2,X=3\}+P\{Y=2,X=4\}$

$=P\{Y=2|X=2\}P\{X=2\}+P\{Y=2|X=3\}P\{X=3\}$

$+P\{Y=2|X=4\}P\{X=4\}=\frac{1}{2}\times\frac{1}{4}+\frac{1}{4}\times\frac{1}{3}+\frac{1}{4}\times\frac{1}{4}=\frac{13}{48}.$

2.10. (1) $\lim\limits_{x\to(-1)_+} F(x)=b-a=\frac{1}{8}$,

$F(1)-\lim\limits_{x\to 1_-} F(x)=1-(a+b)=\frac{1}{4}\Rightarrow a=\frac{5}{16}, b=\frac{7}{16}$.

$P\{X<1\}=F(1)-P(X=1)=\frac{3}{4}$,

$P(-1\leqslant X\leqslant 1)=P(X\leqslant 1)-P(X<-1)=1$.

2.11. $P(0<X<2)=F(2)-F(0)=1-\frac{1}{2}e^{-2}-\left(1-\frac{1}{2}e^{0}\right)=\frac{1}{2}-\frac{1}{2}e^{-2}$.

2.12. $F(5)=\int_{-\infty}^{5} f(x)\mathrm{d}x=\int_{0}^{5} 2e^{-2x}\mathrm{d}x=1-e^{-10}$.

2.13. $\lim\limits_{x\to 2_+} F(x)=\lim\limits_{x\to 2_+} ax^2=4a=1\Rightarrow a=\frac{1}{4}$.

2.14. $F(x)=\begin{cases} 0, & x\leqslant 0; \\ \frac{1}{2}-\frac{1}{2}\cos x, & 0<x<\pi; \\ 1, & \pi\leqslant x. \end{cases}$

2.15. If $t>0, P(T_1\leqslant t)=1-P(T_1>t)=1-P(N(t)=0)=1-e^{-\lambda t}$. Otherwise, $P(T_1\leqslant t)=0$.

Then the p. d. f. of T_1 is $f(x)=\begin{cases} \lambda e^{-\lambda x}, & x>0; \\ 0, & x\leqslant 0. \end{cases}$

2.16. A: the bolt is qualified.

$P(\overline{A})=1-P(A)=1-\left[\Phi\left(\frac{10+0.12-10}{0.06}\right)-\Phi\left(\frac{10-0.12-10}{0.06}\right)\right]=2-2\Phi(2)$.

2.17. $f(x)=\begin{cases} e^{-x}, & 0\leqslant x; \\ 0, & x<0. \end{cases}$

$Y=X^2$. If $y<0, P(Y\leqslant y)=0$.

Otherwise, $P(Y\leqslant y)=P(-\sqrt{y}\leqslant X\leqslant\sqrt{y})=\int_0^{\sqrt{y}} e^{-x}\mathrm{d}x=1-e^{-\sqrt{y}}$.

$$f_Y(y)=\begin{cases}\dfrac{e^{-\sqrt{y}}}{2\sqrt{y}}, & y>0;\\ 0, & \text{otherwise.}\end{cases}$$

2.18. $X\sim N(2,4), Y=-X+1\sim N(-1,4)$.

2.19. (1) If $y\geqslant 0, P(Y\leqslant y)=P(|X|\leqslant y)=\int_{-y}^{y}\frac{1}{2}e^{-|x|}dx=1-e^{-x}$.

Otherwise, $P(Y\leqslant y)=0$. Then the p. d. f. of Y is $f_Y(y)=\begin{cases}e^{-y}, & y\geqslant 0;\\ 0, & y<0.\end{cases}$

(2) $P(Y\leqslant y)=P(\log|X|\leqslant y)=P(-e^y\leqslant X\leqslant e^y)=\int_{-e^y}^{e^y}\frac{1}{2}e^{-|x|}dx=$
$1-e^{-e^y}$.

Then the p. d. f. of Y is $f_Y(y)=e^{y-e^y}$.

2.20. If $y\leqslant 0$, $f_Y(y)=0$.

Otherwise, $y>0$, $f_Y(y)=\frac{1}{\sqrt{2\pi y}}e^{-\frac{(\ln y)^2}{2}}$.

Chapter 3

3.1. Let X be the number of blue balls taken and Y be the number of red balls taken.

(1) The case for the replacement:

$P(X=0,Y=2)=\frac{1}{36}$,

$P(X=1,Y=1)=\frac{10}{36}=\frac{5}{18}$,

$P(X=2,Y=0)=\frac{25}{36}$.

(2) The case for the without replacement:

$P(X=0,Y=2)=\frac{1}{66}$,

$$P(X=1,Y=1)=\frac{20}{66}=\frac{10}{33},$$

$$P(X=2,Y=0)=\frac{45}{66}=\frac{15}{22}.$$

3.2. Let X be the number of heads and Y be the absolute value of heads and tails.

$$P(X=0,Y=4)=\frac{1}{16},$$

$$P(X=1,Y=2)=\frac{4}{16}=\frac{1}{4},$$

$$P(X=2,Y=0)=\frac{6}{16}=\frac{3}{8},$$

$$P(X=3,Y=2)=\frac{4}{16}=\frac{1}{4},$$

$$P(X=4,Y=0)=\frac{1}{16}.$$

3.3. (1)

X \ Y	1	2	3	4	5	6
1	0	1/9	0	0	0	0
2	0	2/9	1/9	0	0	0
3	0	0	0	2/9	2/9	1/9

(2)

X \ Y	1	2	3
1	1/9	1/9	1/9
2	0	2/9	1/9
3	0	0	3/9

3.4. $P\{X_1=k,X_2=l\}=(1-p)^{k+l}p^2,k,l=1,2,\cdots$

3.5. (1) $k=\frac{1}{8}$.

(2) $\frac{3}{8}, \frac{5}{8}, \frac{5}{8}$.

(3) $f(x)=\begin{cases}\frac{1}{4}(3-x), & 0<x<2;\\ 0, & \text{otherwise.}\end{cases}$

$f(y)=\begin{cases}\frac{1}{4}(5-y), & 2<y<4;\\ 0, & \text{otherwise.}\end{cases}$

For $2<y<4, f_{X|Y}(x|y)=\begin{cases}\frac{6-x-y}{2(5-y)}, & 0<x<2;\\ 0, & \text{otherwise.}\end{cases}$

For $0<x<2, f_{Y|X}(y|x)=\begin{cases}\frac{6-x-y}{2(3-x)}, & 2<y<4;\\ 0, & \text{otherwise.}\end{cases}$

3.6. (1) $k=2, \frac{e^2-1}{e^3}$.

(2) $\frac{1}{3}$.

(3) $1-e^{-a}$.

(4) $f(x)=\begin{cases}e^{-x}, & x>0;\\ 0, & \text{otherwise.}\end{cases}$ $f(y)=\begin{cases}2e^{-2y}, & y>0;\\ 0, & \text{otherwise.}\end{cases}$

当 $y>0$ 时，$f_{X|Y}(x|y)=\begin{cases}e^{-x}, & x>0;\\ 0, & \text{otherwise.}\end{cases}$

当 $x>0$ 时，$f_{Y|X}(y|x)=\begin{cases}2e^{-2y}, & y>0;\\ 0, & \text{otherwise.}\end{cases}$

(5) Independent.

3.7. (1) $c=\frac{1}{\pi}$.

(2) $f(x)=\begin{cases}\frac{2}{\pi}\sqrt{1-x^2}, & -1\leqslant x\leqslant 1;\\ 0, & \text{otherwise.}\end{cases}$

$$f(y)=\begin{cases}\frac{2}{\pi}\sqrt{1-y^2}, & -1\leqslant y\leqslant 1;\\ 0, & \text{otherwise.}\end{cases}$$

(3) $\frac{1}{\pi a^2}$.

(4) $$f(x)=\begin{cases}\frac{2}{\pi a^2}\sqrt{a^2-x^2}, & -a\leqslant x\leqslant a;\\ 0, & \text{otherwise.}\end{cases}$$

$$f(y)=\begin{cases}\frac{2}{\pi a^2}\sqrt{a^2-y^2}, & -a\leqslant y\leqslant a;\\ 0, & \text{otherwise.}\end{cases}$$

3.8. omitted.

3.9.

For $-1<y\leqslant 0$, $f_{X|Y}(x|y)=\begin{cases}\frac{1}{1+y}, & -y\leqslant x\leqslant 1;\\ 0, & \text{otherwise.}\end{cases}$

For $0\leqslant y<1$, $f_{X|Y}(x|y)=\begin{cases}\frac{1}{1-y}, & y\leqslant x\leqslant 1;\\ 0, & \text{otherwise.}\end{cases}$

For $0<x\leqslant 1$, $f_{Y|X}(y|x)=\begin{cases}\frac{1}{2x}, & |y|<x;\\ 0, & \text{otherwise.}\end{cases}$

3.10.

(1) $\frac{5}{8}$.

(2) $f(x)=\begin{cases}-3x^2+2x+1, & -1\leqslant x\leqslant 1;\\ 0, & \text{otherwise.}\end{cases}$

(3) For $-1\leqslant x\leqslant 1$, $f_{Y|X}(y|x)=\begin{cases}\frac{2(x+y)}{-3x^2+2x+1}, & x\leqslant y\leqslant 1;\\ 0, & \text{otherwise.}\end{cases}$

3.11.

(1) $f(y)=\begin{cases} -9y^2\ln y, & 0<y<1; \\ 0, & \text{otherwise}. \end{cases}$

(2) For $0<y<1$, $f_{X|Y}(x|y)=\begin{cases} -\dfrac{1}{x\ln y}, & y<x<1; \\ 0, & \text{otherwise}. \end{cases}$

3.12.

(1) Independent.

(2) Not independent.

3.13. $\dfrac{2}{3}$.

3.14. Omitted.

3.15.

(1) $f(x,y)=\begin{cases} \dfrac{1}{2}e^{-\frac{y}{2}}, & 0<x<1, y>0; \\ 0, & \text{otherwise}. \end{cases}$

(2) $1-\int_0^1 e^{-\frac{x^2}{2}}dx$.

3.16.

$P(Z=0)=\dfrac{1}{3}$, $P(Z=1)=\dfrac{2}{3}$, $Z\sim B\left(1,\dfrac{2}{3}\right)$.

3.17.

X \ Y	0	1	2	Total
0	1/6	1/12	1/12	1/3
1	1/3	1/6	1/6	2/3
Total	1/2	1/4	1/4	1

Z_1	−1	0	1	2
P	1/3	1/3	1/4	1/12

Z_2	0	1	2	3
P	1/6	5/12	1/4	1/6

3.18. Omitted.

3.19. $P(Z=k)=(k-1)p^2(1-p)^{k-2}, \quad k=2,3,\cdots$

3.20. Omitted.

3.21.

$$f_Z(z)=\begin{cases} z, & 0<z<1; \\ 2-z, & 1\leqslant z<2; \\ 0, & \text{otherwise.} \end{cases} \qquad f_{\frac{z}{2}}(z)=\begin{cases} 4z, & 0<z<2; \\ 4-4z, & 2\leqslant z<4; \\ 0, & \text{otherwise.} \end{cases}$$

3.22.

$$f_Z(z)=\begin{cases} z^2, & 0<z<1; \\ 2z-z^2, & 1\leqslant z<2; \\ 0, & \text{otherwise.} \end{cases}$$

3.23.

(1) $f_Z(z)=\begin{cases} ze^{-z}, & z>0; \\ 0, & \text{otherwise.} \end{cases}$

(2) $f_{X-Y}(z)=\begin{cases} \dfrac{1}{2}e^{-z}, & z>0; \\ 0, & \text{otherwise.} \end{cases}$

(3) $f_{\frac{X}{Y}}(z)=\begin{cases} \dfrac{1}{(1+z)^2}, & z>0; \\ 0, & \text{otherwise.} \end{cases}$

3.24. $f_Z(z)=\begin{cases} 1-e^{-z}, & z>0; \\ 0, & \text{otherwise.} \end{cases}$

3.25.

(1) not independent

(2) $f_Z(z)=\begin{cases} \dfrac{1}{2}z^2e^{-z}, & z>0; \\ 0, & \text{otherwise.} \end{cases}$

3.26. $f_Z(z)=\begin{cases} \dfrac{2z}{1+z^2}, & z\geqslant 0; \\ 0, & \text{otherwise.} \end{cases}$

3.27. Omitted.

3.28.

(1) 1.

(2) $f(x)=\begin{cases} e^{-x}, & x>0; \\ 0, & \text{otherwise.} \end{cases}$ $f(y)=\begin{cases} e^{-y}, & y>0; \\ 0, & \text{otherwise.} \end{cases}$

(3) $f_U(u)=\begin{cases} 2(1-e^{-u})e^{-u}, & u>0; \\ 0, & \text{otherwise.} \end{cases}$

$f_V(v)=\begin{cases} 2e^{-2v}, & v>0; \\ 0, & \text{otherwise.} \end{cases}$

3.29.

$$F_M(m)=\begin{cases} 0, & m\leqslant 0; \\ 1-(1-m)^n, & 0<m<1; \\ 1, & m\geqslant 1. \end{cases}$$

$$F_N(y)=\begin{cases} 0, & y<0; \\ y^n, & 0\leqslant y<1; \\ 1, & y\geqslant 1. \end{cases}$$

3.30. $f(u,v)=\frac{1}{4\pi}e^{-\frac{u^2+v^2}{4}}$.

3.31. Omitted.

Chapter 4

4.1. $-\frac{1}{4}$ $\frac{3}{4}$ $\frac{9}{4}$.

4.2. 14.

4.3. $\frac{1}{3}$.

4.4. $2\sigma^2$.

4.5. Omitted.

4.6. Omitted.

4.7. Omitted.

4.8. (1) $f(y)=\begin{cases} e^{-y}, & y>0; \\ 0, & \text{otherwise}. \end{cases}$ $E(Y)=1$.

(2) $E(Y)=1$.

4.9. $E(U)=2, E(V)=1/2, E(W)=1/2$.

4.10. Notice that (1) X has the p. f. $P\{X=k\}=(\frac{1}{6})^{k-1}(1-\frac{1}{6}), k=1,2,\cdots$, and (2) X and Y are independent. So, $E(X)=E(X|Y=1)=E(X|Y=5)=6$.

4.11. (1) $\frac{y^3}{4}$. (2) $2y^2$.

4.12. $E(X)=10p, \mathrm{Var}(X)=10p(1-p)$.

4.13. $E(X)=\frac{1}{p}, \mathrm{Var}(X)=\frac{1-p}{p^2}$.

4.14. Omitted.

4.15. (1) $E(X)=2, E(Y)=0, \mathrm{Var}(X)=0.8, \mathrm{Var}(Y)=0.6$.

(2) $E\left(\frac{Y}{X}\right)=-\frac{1}{15}$.

(3) $E(X-Y)^2=5$.

4.16. (1) $E(X)=1/6, \mathrm{Var}(X)=5/36$.

(2) $E(X)=5/3, \mathrm{Var}(X)=25/18$.

4.17. $E(X)=Np, \mathrm{Var}(X)=Np(1-p)$.

4.18 (1) $P\{X=k\}=C_{k-1}^{r-1}p^r(1-p)^{k-r}, k=r, r+1, \cdots$

(2) $E(X)=\frac{r}{p}, \mathrm{Var}(X)=\frac{r(1-p)}{p^2}$.

4.19. $\rho_{XY}=-1/5$.

4.20. 0.

4.21. $\rho(Y_n, Y_{n+j})=\begin{cases} \frac{1}{2}, & j=1, \\ 0, & j>1. \end{cases}$

4.22. (1) 1/2, (2) 0.

4.23. Omitted.

4.24. Uncorrelated and not independent.

4.25. $-\frac{3}{2}$.

4.26. $\frac{a^2-b^2}{a^2+b^2}$.

4.27. $-\frac{1}{11}$.

4.28. (1) 10, (2) 31.

4.29. 3.

4.30. $\begin{pmatrix} \frac{13}{162} & -\frac{1}{81} \\ -\frac{1}{81} & \frac{23}{81} \end{pmatrix}$.

4.31. $\begin{pmatrix} U \\ V \\ W \end{pmatrix} = \begin{pmatrix} 2 & 1 & 3 \\ 1 & -2 & 4 \\ 3 & 4 & 1 \end{pmatrix} \begin{pmatrix} X \\ Y \\ Z \end{pmatrix}$, $\boldsymbol{C}_{UVW} = \boldsymbol{A}\boldsymbol{C}_{XYZ}\boldsymbol{A}^{\mathrm{T}} = \begin{pmatrix} 86 & 53 & 100 \\ 53 & 44 & 45 \\ 100 & 45 & 140 \end{pmatrix}$.

4.32. $b=\frac{\sigma_1}{\sigma_2}$.

4.33. (1) not independent, (2) independent.

Chapter 5

5.1. Omitted.

5.2. To prove by Chebyshev's inequality.

5.3. 0,1.

5.4. 0.

5.5. n satisfies $2\Phi\left(\frac{\sqrt{n}}{10}\right)-1\geqslant 0.95$.

5.6. (1) $2\Phi\left(\frac{10}{\sqrt{47.5}}\right)-1$. (2) $1-\Phi\left(\frac{20}{\sqrt{47.5}}\right)$.

5.7. $1-\Phi\left(\frac{35}{4}\right)$.

Chapter 6

6.1. 40.

6.2. (1) 0.262 8. (2) 0.292 3, 0.578 5.

6.3. (1) $P\{X_1=x_1, X_2=x_2, \cdots, X_n=x_n\}=\frac{e^{-n\lambda}\lambda^{\sum_{i=1}^{n}x_i}}{\prod_{i=1}^{n}(x_i!)}, x_i=0,1,$ $2\cdots(i=1,2,\cdots,n)$

(2) $\lambda, \frac{\lambda}{n}, \lambda$.

6.4. $2(n-1)\sigma^2$.

6.5. 0.829 3.

6.6. χ^2 distribution with one degree of freedom.

6.7. $c=\frac{1}{3}$.

6.8. $c=\sqrt{\frac{3}{2}}$.

6.9. Proof omitted.

6.10. $t(n-1)$.

6.11. (1) $P\{X_1=x_1, X_2=x_2, \cdots, X_n=x_n\}=p^{\sum_{i=1}^{n}x_i}(1-p)^{n-\sum_{i=1}^{n}x_i}$.

(2) $\binom{n}{k}p^k(1-p)^{n-k}, k=0,1,2,\cdots,n$.

(3) $p, \frac{1}{n}p(1-p), p(1-p)$.

6.12. (1) $f(x_1,\cdots,x_{10}) = \dfrac{1}{(2\pi\sigma^2)^5}\mathrm{e}^{-\sum\limits_{i=1}^{10}(x_i-\mu)^2/(2\sigma^2)}$.

(2) $f_{\bar{x}}(x)=\dfrac{\sqrt{5}}{\sqrt{\pi}\sigma}\mathrm{e}^{-5(x-\mu)^2/\sigma^2}$.

6.13. (1) $f(x)=\begin{cases}\dfrac{1}{4.32}\left(\dfrac{x}{0.18}\right)^4\mathrm{e}^{-\frac{x}{0.18}}, & x>0;\\ 0, & x\leqslant 0.\end{cases}$

(2) 0.1.

6.14. $f(x)=\begin{cases}\dfrac{n\lambda^n}{(n-1)!}(nx)^{n-1}\mathrm{e}^{-n\lambda x}, & x>0;\\ 0, & x\leqslant 0.\end{cases}$

6.15. (1) $p=0.99$. (2) $\dfrac{2}{15}\sigma^4$.

Chapter 7

7.1. (1) $\hat{\mu}=74.002$, $\hat{\sigma}^2=6\times10^{-6}$. (2) $s^2=6.86\times10^{-6}$.

7.2. (1) $\hat{\theta}=\dfrac{\overline{X}}{\overline{X}-c}$. (2) $\hat{\theta}=(\dfrac{\overline{X}}{1-\overline{X}})^2$. (3) $\hat{p}=\dfrac{\overline{X}}{m}$.

7.3. (1) $\hat{\theta}=\dfrac{1}{(\frac{1}{n}\sum\limits_{i=1}^{n}\ln X_i-\ln c)}$. (2) $\hat{\theta}=\dfrac{n^2}{(\sum\limits_{i=1}^{n}\ln X_i)}$. (3) $\hat{p}=\dfrac{\overline{X}}{m}$.

7.4. Proof omitted.

7.5. (1) $\hat{\theta}=\overline{X}$. (2) Proof omitted.

7.6. $\hat{\sigma}^2=\dfrac{1}{n}\sum\limits_{i=1}^{n}(X_i-\overline{X})^2$.

7.7. $\hat{\beta}=\dfrac{1}{\overline{X}}$.

7.8. Proof omitted.

7.9. $\hat{\theta}=\overline{X}$.

7.10. $\hat{\theta}_1=\min(X_1,\cdots,X_n)$; $\hat{\theta}_2=\max(X_1,\cdots,X_n)$.

7.11. $\frac{5}{6};\frac{5}{6}$.

7.12. $\hat{\lambda}=\overline{X},\hat{\lambda}=\overline{X}$.

7.13. Proof omitted.

7.14. (1) $\hat{\theta}=-\frac{1}{n}\sum_{i=1}^{n}X_i$. (2) Proof omitted.

7.15. (1) T_1,T_3. (2) T_3 is more effective than T_1.

7.16. (1) Proof omitted. (2) $\left(\frac{\chi^2_{1-\alpha/2}(2n)}{2n\overline{X}},\frac{\chi^2_{\alpha/2}(2n)}{2n\overline{X}}\right)$. (3) $\frac{\chi^2_{\alpha}(2n)}{2n\overline{X}}$.

7.17. (1) (2.719 2, 3.405 8). (2) (2.634 3, 3.490 7).
(3) (2.428 4, 3.696 6).

7.18. (148.1, 165.6)

7.19. The confidence interval for σ^2 is $\left(\frac{\sum_{i=1}^{n}(X_i-\mu)^2}{\chi^2_{\alpha/2}(n)},\frac{\sum_{i=1}^{n}(X_i-\mu)^2}{\chi^2_{1-\alpha/2}(n)}\right)=$ (5.013,31.626).

The confidence interval for σ is (2.239,5.624).

7.20. (1) (−0.002, 0.006). (2) −0.0012.

7.21. (1) (0.222,3.601). (2) 2.84.

Chapter 8

8.1. Proof omitted.

8.2. Proof omitted.

8.3. (1) 0.03. (2) 0.003. (3) 0.004.

8.4. (U-test) Accept the Null hypotheses. ($H_0:\mu=\mu_0$)

8.5. (t-test) Accept the Null hypotheses.

8.6. (paired-test) Accept the Null hypotheses.

8.7. (t-test) Reject the Null hypotheses. ($H_0: \mu_0 = 100$)

8.8. (F-test) Accept the Null hypotheses.

8.9. (chi-test) Reject the Null hypotheses. ($H_0: \sigma_0 = 14$)

8.10. (F-test) Accept the Null hypotheses.

8.11 Proof omitted.

8.12 (1) Proof omitted. (2) Yes.

8.13. Accept the Null hypotheses.

8.14. (1) NO. (2) NO. (3) YES.

8.15. (t-test) Accept the Null hypotheses. ($H_0: \mu_0 = 5$)

Chapter 9

9.1. Proof omitted.

9.2. Proof omitted.

9.3. Proof omitted.

9.4. (a) $y = 40.893 + 0.548x$. (b) $38.483 + 3.440x - 0.643x\hat{}2$.

9.5. Proof omitted.

9.6. Proof omitted.

9.7. Proof omitted.

9.8. Proof omitted.

9.9. (1) $\hat{y} = -0.104 + 0.988x$. (2) (13.29, 14.17).

9.10. (1) $\hat{y} = 13.9584 + 12.5503x$. (2) Rejected H_0. (3) (11.82, 13.28).

References

[1] Deep R. Probability and Statistics [M]. New York Academic Press, 2005.

[2] DevoreJay L. The Probability and Statistics for Engineering and the Sciences [M]. Cambridge Wadsworth Group Press, 2004.

[3] Gyenis Z, Rédei M. Defusing Bertrand's paradox [J]. British Journal for the Philosophy of Science, 2014, 65:349-373.

[4] Hogg Robert V, Craig Allen T. Introduction to Mathematical Statistics [M]. 5th ed. Beijing: Higher Education Press, 2012.

[5] Iversen G R. 统计学 [M]. 吴喜之,等,译. 北京:高等教育出版社, 2000.

[6] Jacod J, Protter P. Probability Essentials [M]. 2nd ed. Springer, 2002.

[7] Kallenberg O. Foundations of Modern Probability [M]. 2nd ed. Springer, 2002.

[8] Ross S M. Introduction to Probability Models [M]. 6th ed. New York Academic Press, 1997.

[9] Ross S M. A First Course in Probability [M]. 7th ed. New Jersey, USA Prentice Hall Press, 2005.

[10] 邓集贤,杨维权,司徒荣,等. 概率论及数理统计 [M]. 4 版. 北京:高等教育出版社, 2009.

[11] 房祥忠,鲁立刚,李东风改编. 概率论与数理统计(改编版,原著 DeGroot M H, Schervish M J) [M]. 3 版. 北京:高等教育出版社, 2005.

[12] 胡细宝,孙洪祥,王丽霞. 概率论·数理统计·随机过程[M]. 北京:北京邮电大学出版社, 2004.

[13] 茆诗松，程依明，濮晓龙. 概率论与数理统计教程[M]. 北京：高等教育出版社，2004.
[14] 盛骤，谢式千，潘承毅. 概率论与数理统计[M]. 3 版. 北京：高等教育出版社，2005.
[15] 庄楚强，等. 应用数理统计基础[M]. 广州：华南理工大学出版社，2002.